DISCARDED

JUN 2 6 2025

Asheville-Buncombe
Technical Community College
Learning Resources Center
340 Victoria Road
Asheville, NC 28801

CRYOGENIC TECHNOLOGY
AND APPLICATIONS

CRYOGENIC TECHNOLOGY AND APPLICATIONS

AMSTERDAM • BOSTON • HEIDELBERG • LONDON
NEW YORK • OXFORD • PARIS • SAN DIEGO
SAN FRANCISCO • SINGAPORE • SYDNEY • TOKYO

Butterworth–Heinemann is an imprint of Elsevier
30 Corporate Drive, Suite 400, Burlington, MA 01803, USA
Linacre House, Jordan Hill, Oxford OX2 8DP, UK

Copyright © 2006, Elsevier Inc. All rights reserved.

No part of this publication may be reproduced, stored in a retrieval system, or transmitted in any form or by any means, electronic, mechanical, photocopying, recording, or otherwise, without the prior written permission of the publisher.

Permissions may be sought directly from Elsevier's Science & Technology Rights Department in Oxford, UK: phone: (+44) 1865 843830, fax: (+44) 1865 853333, E-mail: permissions@elsevier.com. You may also complete your request on-line via the Elsevier homepage (http://elsevier.com), by selecting "Support & Contact" then "Copyright and Permission" and then "Obtaining Permissions."

Recognizing the importance of preserving what has been written, Elsevier prints its books on acid-free paper whenever possible.

Library of Congress Cataloging-in-Publication Data
Application Submitted

British Library Cataloguing-in-Publication Data
A catalogue record for this book is available from the British Library.

ISBN 13: 978-07506-7887-2
ISBN 10: 0-7506-7887-9

For information on all Butterworth–Heinemann publications
visit our Web site at www.books.elsevier.com

Printed in the United States of America
05 06 07 08 09 10 10 9 8 7 6 5 4 3 2 1

Working together to grow
libraries in developing countries

www.elsevier.com | www.bookaid.org | www.sabre.org

ELSEVIER   BOOK AID International   Sabre Foundation

Dedication

This book is dedicated to my beloved parents who always encouraged me to pursue advanced research studies in the fields of science and technology for the benefit to mankind.

Foreword

This book comes at a time when the future well-being of nations in the 21st century's global economy increasingly depends on the quality and depth of the technological innovations they can commercialize rather than on the "old money" they may have inherited through historical conquests or pure luck bestowed by natural resources. The book deals with just about all aspects of cryogenic technology, which is often the "hidden" enabler of many wide-ranging modern technological marvels — ranging from superconducting electromagnets employed to levitate ultra-high-speed trains in Shanghai to ultrahigh purity metallic nano-technology polyamide vapor deposition (PVD) films deposited at 10E-9 torr made possible by cryopumps to the now ubiquitous noninvasive medical diagnostic equipment such as the magnetic resonance imaging (MRI) and computer tomography (CT) scans.

The author of this book, Dr. A. R. Jha, continues his distinguished track record of distilling complex theoretical physics concepts into an understandable technical framework suitable for extending to practical applications across a wide front of modern industries. I found particularly refreshing his big-picture approach, without compromising on the basic science behind it all. This should help present-day students, both undergraduate and graduate, master difficult scientific concepts with full confidence that they will have commercial/engineering applications to benefit the emerging economies across the world. The book is well organized, does not assume a prior familiarity with cryogenics, and derives relevant mathematical expressions from first principles.

For example, the book provides a treatment of the underlying thermodynamics of cooling systems illustrated by temperature-entropy diagrams, develops nonlinear heat flow equations, and summarizes new solid-state coolants such as rare earths that provide unique capabilities now available for a variety of applications. It then goes on to describe the practical systems that have evolved since the mid-1960s — ranging from the ubiquitous home refrigerator to a variety of cryocoolers capable of temperatures ranging from liquid CO_2 → liquid N_2 → liquid He and beyond, thanks to pumping capabilities that can be combined therein. Of particular interest for portable applications are microcoolers that can sustain uniform temperature gradients with high-efficiency heat exchangers. At a practical level, the author treats the important end-user issues relating to reliability, power consumption, and other factors.

A variety of miniature electronic devices achieve higher, more useful performance levels when cooled down to lower temperatures with the microcoolers. Present-day microprocessors and logic devices for high-speed servers and mainframe computers are running into severe power dissipation limitations, which can be overcome by using microcoolers to extend their performance curve. Josephson tunneling junction-based devices that only operate at lower temperatures may one day become practical with commercial availability of suitable coolers. At portable levels, microcoolers enable infrared imaging sensors, high-resolution charge-coupled devices (CCDs) electrooptic devices, and mm wave airborne radiometers. This book does a good job of explaining these basic applications enabled through cryogenics.

The author's personal background allows him to provide an authoritative account of many emerging applications of small size, high-reliability cryogenic devices for portable applications such as sensors in unmanned aerial vehicles, space-based pollution measurement systems, laser-based surveillance, and portable military electronics utilizing infrared (IR) imaging.

In the realm of moderate and high-capacity cryogenics applications, the author elucidates the technology behind continuing advances in important, well-known industrial applications such as cryogenically cooled turbo-molecular pumps, high-capacity mechanical refrigerators, and high-pressure gas coolers. He covers practical aspects of these applications — such as reliability and ease of maintenance — that impact the overall cost of ownership. Continued competitiveness in such high-volume, potentially high-profit industries requires continuous improvement programs, which will benefit from exposure to the broad context this book provides the reader.

Then there are many exciting, high-profile, enabling applications made possible by ultra-high capacity coolers. Medical diagnostic equipment such as MRI systems and CT scanners need the right kind of cryogenics to make them functional. Superconducting magnet-based systems such as linear accelerators for subatomic particle research and high-speed levitation transportation systems have a common need for capable cryogenics in order to be effective. Other emerging applications include ultrahigh-capacity sonar systems for underwater detection and new chemical oxygen iodine lasers (COIL).

A wide variety of readers should benefit from this book. Clearly the author has in mind advanced undergraduate and graduate students of mechanical and materials engineering who might wish to pursue a career in the cryogenics industry. However, in view of critical interdependencies with other technical disciplines, this book also should be of interest to a wider variety of engineering students or practicing engineers in industries such as medical equipment, defense electronics, security, space, and other yet-to-be-established disciplines. It should be particularly useful for industrial research engineers interested in self-study to broaden their competencies and combat technical obsolescence, since the author has presented an updated view of this entire field. Technical managers should find this book useful to help them update their applications know-how and gain ideas to expand the business opportunities of their companies.

In conclusion, I strongly recommend this book to the broad audience including students, project managers, research engineers, life-science scientists, clinical scientists, and project engineers deeply involved in the design and development of cryogenically cooled sensors and systems best suited for industrial, commercial, military, and space applications.

Dr. A. K. Sinha
Sr. Vice President
Applied Materials, Inc.
Santa Clara, California (United States)

Preface

Advancements in the cryogenic technology and state-of-the art cryocooler designs have opened the door for potential applications in cryogenically cooled sensors and devices such as photonic equipment, electro-optic components, and high-resolution infrared sensors best suited for commercial, industrial, military, and space applications. Cryogenically cooled photonic, electrooptic, microwave millimeter-wave (MM-wave), and infrared (IR) devices have potential applications in space surveillance and reconnaissance, premises security, missile warning, and medical diagnosis and treatment. Cryogenic cooling has demonstrated the significant performance improvement of sensors deployed by unmanned aerial vehicles (UAVs), IR search and tracking systems, radar and missile warning receivers, satellite tracking systems, space-based pollution monitoring equipment, IR imaging sensors using focal planar array technology, forward looking radar and IR systems, high-power and laser-based surveillance systems, space-based multispectral and hyperspectral sensors, and a host of other commercial, industrial, and military systems. Cryogenic technology plays a key role in magnetic resonance imaging (MRI) and computer tomography (CT) systems capable of providing ultrahigh resolution images vital for medical diagnosis.

This book summarizes important aspects of cryogenic technology critical to the design and development of several refrigerators, cryocoolers, and microcoolers deployed by various commercial, industrial, space, and military systems. It is important to mention that the integration of cryogenic technology in electrooptic, microwave, optoelectronic, MM-wave, photonic, and IR devices and systems is necessary if optimum performance, improved reliability, and long operating life are the critical requirements. Performance improvements at cryogenic temperatures in various devices and systems are identified with emphasis on reliability and power consumption. This book presents a balanced mix of theory and practical applications. Mathematical expressions and their derivations highlighting performance enhancement are provided for the benefit of students who intend to pursue higher studies in cryogenic engineering. This book is well organized and covers critical design aspects representing cutting-edge cryogenic technology. It is written in a language most beneficial to undergraduate and graduate students who are willing to expand their horizons in cryogenic technology. The integration of cryogenic technology with monolithic microwave and integrated circuit (MMIC) technology and surface mount technique is identified to achieve enhanced performance, improved reliability, and minimum power consumption.

This book has been written specially for cryogenic engineers, research scientists, professors, project managers, educators, clinical researchers, and program managers deeply engaged in the design, development, and research of cryogenically cooled devices and systems for various applications. The book will be most useful to those who wish to broaden their knowledge of cryogenic engineering. The author has made every attempt to provide well-organized material using conventional nomenclature, a consistent set of symbols, and identical units so that readers with little knowledge in the field will be able to rapidly comprehend the material. The latest performance parameters and experimental data on cryogenically cooled devices and sensors are provided in this book, which have been taken from various references with due credits to authors and sources. The bibliographies provided include significant contributing sources. This book is composed of eight chapters, each dedicated to a specific topic and application.

Chapter 1 summarizes the earlier scientific developments and technological advancements in cryogenic engineering and technology. The chapter focuses on the anticipated benefits of cryogenic technology and briefly summarizes the performance of various cooling systems operating since the mid-1960s.

Chapter 2 identifies the continuing advances in the cryogenically cooled electronics such as Josephson Junction (JJ) devices that will ultimately replace the traditional semiconductor devices currently used in digital circuit applications. The integration of cryogenic technology with the next-generation, high-performance semiconductors will offer significant improvements in speed, size, power consumption, and reliability of the device or sensor.

Chapter 3 focuses on the thermodynamic aspects of various cooling systems, with an emphasis on physical parameters such as the temperature-entropy diagram, constant enthalpy, heat leak, and cooling load. The impact of heat flow on the heat transfer efficiency of the heat exchanger is identified. The dependence of second-order nonlinear heat flow equations on cooler capacity and input power requirements is investigated.

Chapter 4 identifies the microcooler technology best suited for space sensors, satellite communications involving cryoelectronics, and complex military systems where cooler size, efficiency, reliability, and power consumption are of critical importance. Requirements for microcoolers for high-resolution imaging sensors operating under harsh environments are defined, with an emphasis on reliability and maintenance. Microcoolers capable of maintaining uniform-temperature gradients requiring high heat exchanger efficiency are briefly discussed.

Chapter 5 defines performance requirements for moderate- and high-capacity cryocoolers. Such coolers include high-pressure gas coolers, J-T cooling systems, cryogenic turbo-pumps, and high-capacity mechanical refrigerators. Performance capabilities and limitations of these coolers are described in terms of logistics, cooling cycles, reliability, and maintenance procedures. The impact of the high-pressure ratio on energy efficiency, operating cost, and coolant-mass flow is identified.

Chapter 6 identifies the performance requirements of microcoolers best suited for airborne military systems and space sensors. Cryogenic cooling of sensors and devices such as infrared imaging sensors, IR detectors, IR focal planar arrays, high-resolution CCD-based cameras, IR line scanners, IR search and track sensors, and MM-wave airborne radiometers is absolutely necessary to achieve state-of-the art performance.

Chapter 7 describes ultrahigh capacity cooling systems best suited for sonar transmitters capable of providing underwater detection, mapping systems needed for naval warfare applications, medical diagnostic equipment, and propulsion systems. High capacity and high efficiency are required for superconducting linear accelerators; fusion energy test facility; superconducting electrical machines such as motors, generators, and propulsion systems; superconducting bearings; medical diagnosis equipment such as MRI and CT; and megawatt-class airborne laser systems such as the chemical oxygen iodine laser (COIL) capable of providing effective terminal defense against long-range hostile missiles. How cryogenic technology played a key role in the design and development of high-speed trains traveling at speeds exceeding 330 km per hour is briefly discussed.

Chapter 8 defines the requirements for the cryogenic materials and coolants needed by cooling systems to meet their specific performance goals. Rare-earth materials capable of achieving higher cooling efficiency and lower cooling temperatures are identified. Operational problems generally observed in cooling systems such as moisture condensation, freezing of pipes carrying the refrigerants, changes in dimensions with temperature, variation in thermal conductivity of critical cooler elements, leakage of coolants, and inadequate maintenance procedures are briefly discussed. Techniques are identified that will minimize the impact of these problems on the overall system performance without compromising cost or reliability.

I wish to thank Joel Stein, senior editor, for mechanical and material engineering at Elsevier Science, who has been very patient in accommodating last-minute additions and changes to the text. His suggestions have helped the author to prepare the manuscript with remarkable coherency. I also wish to thank my wife, Urmila Jha (who has been patient and supportive throughout the preparation of this book), daughter, Sarita Jha, and my son-in-law, Anu Manglani, who inspired me to complete the book on time under a tight schedule.

Dr. A. R. Jha, Senior Consulting Scientist

Contents

Chapter 1 Technology Advancements and Chronological Development History of Cryogenic Technology   1

- 1.0 Introduction   1
- 1.1 Terms and Phenomena Associated with Cryogenic Systems   2
- 1.2 Prominent Contributors to Cryogenic Technology   2
- 1.3 Critical Aspects and Issues Involved in Cryogenics   5
- 1.4 Benefits from Integration of Cryogenic Technology   6
  - 1.4.1 Affordability   6
  - 1.4.2 Availability   7
  - 1.4.3 Performance Improvement in Various Components and Systems   7
- 1.5 Early Applications of Cryogenic Technology   8
  - 1.5.1 Cryogenic Technology for the Production of Gases   8
  - 1.5.2 Cryogenic Technology for the Production of Inert Gases   8
  - 1.5.3 Cryogenic Technology for Aerospace Applications   8
  - 1.5.4 Cryogenic Liquid Level Controller (LLC)   9
  - 1.5.5 Cryogenic Line Regulators   9
- 1.6 Gas Separation Process Using Cryogenic Technology   10
- 1.7 Industrial Applications of Cryogenic Fluid Technology   10
  - 1.7.1 Liquid Neon   11
  - 1.7.2 Liquid Hydrogen   11
  - 1.7.3 Liquid Nitrogen (LIN)   11
- 1.8 Heat Capacity of Commercial Refrigerants   12
- 1.9 Cryogenic Requirements for the Frozen Food Industry   13
  - 1.9.1 Cold Storage Requirements   13
- 1.10 Cryogenic Requirements for Medical Applications   14
  - 1.10.1 Cryogenic System Requirements for High-Resolution MRI   16
- 1.11 Industrial Applications of Cryogenic Technology   16
  - 1.11.1 Cryopumping   17
  - 1.11.2 Nuclear Radiation Testing   17
  - 1.11.3 Ice-Making Machines and Ice Storage Systems   18
  - 1.11.4 Chilled Water Storage (CWS) Systems   19
- 1.12 Summary   21

xvi   Contents

Chapter 2   Effects of Heat Flow on Heat Exchanger Performance
             and Cooler Efficiency   25

2.0   Introduction   25
2.1   Early Developments in Cryogenic Cooler Technology   26
2.2   Impact of Thermodynamic Aspects on Cryogenic Coolers   27
      2.2.1   Introduction   27
      2.2.2   Symbols and Formulas Widely Used in Thermal Analysis   27
2.3   Types of Heat Flow   28
      2.3.1   Linear Heat Flow   33
              2.3.1.1   Impact of Linear Heat Flow on Heat Exchanger
                        Performance   40
      2.3.2   Impact of Turbulent Heat Flow on Heat Exchanger
              Performance   41
2.4   Two-Dimensional Heat Flow Model   47
      2.4.1   Description of Modified Two-Fluid Model   49
2.5   Heat Transfer Rates for Heat Exchangers   52
      2.5.1   Conduction Mode of Heat Transfer   52
      2.5.2   Convection Mode of Heat Transfer   53
      2.5.3   Radiation Mode of Heat Transfer   54
2.6   Summary   55

Chapter 3   Thermodynamic Aspects and Heat Transfer
             Capabilities of Heat Exchangers for
             High-Capacity Coolers   57

3.0   Introduction   57
3.1   Modes of Heat Transfer Phenomena   57
3.2   Three Distinct Laws of Heat Transfer   58
3.3   Description of Heat Transfer Modes   59
      3.3.1   Conduction   59
      3.3.2   Radiation   60
      3.3.3   Convection   61
3.4   Impact of Heat Transfer Modes on Heat Exchanger Performance   71
      3.4.1   Heat Transfer in a Planar Wall   71
      3.4.2   Heat Transfer in a Composite Wall   75
3.5   Heat Transfer through Heat Exchanger Pipes   76
      3.5.1   Heat Flow in a Cylindrical Pipe   76
      3.5.2   Heat Flow in an Insulated Pipe   78
3.6   Fundamental Design Aspects for a Heat Exchanger   79
      3.6.1   Heat Load Calculations for Heat Exchangers   79
      3.6.2   Computations of Heat Load for the Heat Exchanger (Q_{ex})   83
3.7   Estimates of Heat Removal by Cold Water and Forced Air   85
3.8   Computation of Overall Heat Transfer Coefficient, C_{oht} 86
      3.8.1   Computation of Overall Heat Transfer Coefficient under
              Fouling Conditions   88

		3.8.1.1	Overall Heat Transfer Coefficient for the Foul Condition	88
		3.8.1.2	Overall Heat Transfer Coefficient under Clean Environment	89
3.9	Computation of Critical Parameters for Heat Exchanger			89
	3.9.1	Computation of Temperature Difference for the Countercurrent Flow		90
	3.9.2	Computation of Outside Surface Area (A_{os}) of the Heat Exchanger		91
	3.9.3	Estimation of Heat Transfer Surface Area (A_{os}) under Clean and Fouling Conditions		92
	3.9.4	Computation of Shell Diameter		93
	3.9.5	Computation of Number of Tubes for the Heat Exchanger		93
3.10	Preliminary Rating of a Heat Exchanger			94
3.11	Summary			94

Chapter 4 Critical Design Aspects and Performance Capabilities of Cryocoolers and Microcoolers with Low Cooling Capacities 97

4.0	Introduction		97
4.1	Design Aspects and Operational Requirements		97
4.2	Performance Requirements of Cryocoolers		98
	4.2.1	Maintenance Aspects and Reliability Requirements for Cryocoolers	101
	4.2.2	Cooling Power Requirements for Cryocoolers	102
		4.2.2.1 Cooling Power Requirements for Microcoolers	103
4.3	Cryocoolers Using High-Pressure Ratios		104
	4.3.1	Advantages of High-Pressure Expansion Ratio	106
4.4	Cooling Capacity of a Cryocooler		106
4.5	Temperature Stabilization and Optimization Mass Flow Rate		109
4.6	Advanced Technologies for Integration in Cryocoolers		110
	4.6.1	Pulse Tube Refrigeration (PTR) System Design Aspects and Performance Capabilities	110
4.7	Classifications of Cryocoolers		112
	4.7.1	Stirling-Cycle Cryocoolers	113
	4.7.2	Self-Regulated Joule-Thomson (J-T) Cryocooler	113
	4.7.3	Boreas-Cycle Cryocooler	114
	4.7.4	Closed-Cycle Cryogenic (CCC) Refrigerator	114
	4.7.5	Stirling Cryocooler Using Advanced Technologies	114
	4.7.6	G-M Cryocoolers Employing J-T Valves	118
	4.7.7	G-M-Cycle Cryocoolers	119
	4.7.8	Collin-Cycle Refrigerator Systems	120

Contents

	4.7.9	High-Temperature Refrigerator Systems	120
		4.7.9.1 Cooling Power Requirements at Higher Superconducting Temperatures	122
4.8	Performance Capabilities of Microcoolers		123
	4.8.1	Potential Cooling Schemes for Microcoolers	124
4.9	Performance Capabilities and Limitations of Microcoolers		126
4.10	Specific Weight and Power Estimates for Cryocoolers		127
4.11	Thermodynamic Aspects and Efficiency of Cryocoolers		129
	4.11.1	Thermal Analysis of Refrigeration System	130
4.12	Weight Requirements for Cryogens Used by Cryocoolers		132
4.13	Characteristics and Storage Requirements for Potential Cryogens		133
4.14	Classifications of Cryocoolers and Their Performance Capabilities and Limitations		134
4.15	Summary		134

Chapter 5 Performance Requirements for Moderate- and High-Capacity Refrigeration Systems   139

5.0	Introduction		139
5.1	Description of High-Capacity Refrigeration Systems		139
	5.1.1	Claude-Cycle Refrigeration System	142
	5.1.2	Reversed-Brayton-Cycle Refrigeration System	143
5.2	Refrigeration System with Moderate Cooling Capacity		144
	5.2.1	GM-Cycle Refrigeration System	144
	5.2.2	J-T-Cycle Refrigeration System	146
	5.2.3	Brayton-Cycle Refrigeration System	146
5.3	Turbo-Machinery Refrigeration System		147
5.4	Coefficient of Performance for Various Cooling Systems		147
	5.4.1	Coefficient of Performance for an Ideal Brayton Cooling	149
5.5	Cryogenic Dewar and Storage Tank Requirements for Various Applications		151
5.6	Storage Tank Requirements for Space and Missile Applications		154
	5.6.1	Liquid-Feed Requirements for Storage Systems	155
	5.6.2	Transfer Line Requirements	155
5.7	Operating Pressure and Temperature Requirements for Storage of Liquefied Gases		156
5.8	Cooling System Configurations Using Various Cooling Agents or Cryogens		156
	5.8.1	Characteristics of Various Cryogens	157
	5.8.2	Solid Cryogens	158
	5.8.3	Techniques to Reduce Heat Leak and Weight of Cryogen	160
5.9	Performance Comparison of Various Cryogenic Coolers		160
5.10	Summary		161

Contents   xix

Chapter 6 Cryocooler and Microcoolers Requirements Best Suited for Scientific Research, Military, and Space Applications 165

- 6.0 Introduction 165
- 6.1 Cryocooler Requirements for Various Applications 166
  - 6.1.1 Maintenance Requirements for Cryocoolers 167
- 6.2 Performance Parameters for Various Cryocoolers 168
  - 6.2.1 Dilution-Magnetic Cryocooler 168
  - 6.2.2 Collins-Helium Liquifier (CHL)-Based Cryocooler 168
  - 6.2.3 Gifford-McMahon (G-M) Cryogenic Refrigerator 170
  - 6.2.4 G-M/J-T Refrigerator System 171
  - 6.2.5 Stirling-Cycle Cryocooler 172
  - 6.2.6 Self-Regulated J-T Cryocooler 173
  - 6.2.7 Closed-Cycle, Split-Type (CCST) Stirling Cryocooler 173
- 6.3 Cooling Schemes Used by Various Cooling Systems 177
  - 6.3.1 Choice of Cooling Schemes 177
- 6.4 Microcooler Requirements for Military and Space Applications 178
- 6.5 Cryocoolers Using Unique Design Concepts and Materials 179
  - 6.5.1 Cryocooler and Microcooler Designs Incorporating Rare-Earth Materials 179
  - 6.5.2 Cryocooler Design With High-Pressure Ratio and Counterflow Heat Exchanger 180
- 6.6 Critical Thermodynamic Aspects of Cryocoolers 183
  - 6.6.1 Impact of Thermodynamic Efficiency Limits on Various Cooling Cycles 183
- 6.7 Techniques to Optimize Cooling Capacity 184
- 6.8 Optimization of Temperature Stability and Mass Flow Rate 185
- 6.9 Cryocooler Design Requirements for Space Applications 187
- 6.10 Summary 189

Chapter 7 Integration of Latest Cooler Technologies to Improve Efficiency and Cooling Capacity 191

- 7.0 Introduction 191
- 7.1 Unique Design Concepts and Advanced Materials 191
- 7.2 Design Concepts for a Pulse Tube Refrigerator (PTR) Cryocooler 192
  - 7.2.1 Performance Capabilities of a PTR System 194
  - 7.2.2 Thermodynamic Aspects of a Pulse Tube Cryocooler (PTC) 195
    - 7.2.2.1 Derivation of Expressions for Various Operating Parameters 195
  - 7.2.3 Minimum Refrigeration Temperature (MRT) of a Cryocooler 199

7.3	Ways and Means to Improve PTR Performance		199
	7.3.1 Impact of Coolant and Regenerative Materials on Cooler Performance		202
	7.3.2 Parametric Analysis to Predict PTR System Performance		203
	7.3.3 Impact of Regenerative Materials on Cooler Performance		204
		7.3.3.1 Impact of Phase Shift on Cooling Performance	205
		7.3.3.2 Impact of Heat Leakage on Cooler Efficiency and Cooldown Time	205
7.4	Cryocooler Designs for Industrial Applications		206
	7.4.1 Cooling Power and Adiabatic Efficiency Computations		208
7.5	Multibypass and Active Buffer-Stage Techniques to Improve PTR Cooling Efficiency and Capacity		210
	7.5.1 Implementation of Active-Buffer Stages for Efficiency Enhancement		210
	7.5.2 Integration of Multibypass Technique to Improve Cooling Efficiency		214
7.6	Cryocooler Requirements for Microwave, MM-Wave, and High-Power Laser Systems		215
	7.6.1 Cooler Requirements for Microwave and MM-Wave Systems		215
	7.6.2 Cryogenic Cooler Requirements for Infrared Devices and Sensors		216
	7.6.3 Refrigerator Requirements for High-Power Laser Systems		220
7.7	Cryogenic Coolers for Sonar Applications		221
7.8	Cryogenic Coolers for Medical Applications		222
7.9	Summary		223

Chapter 8 Requirements for Cryogenic Materials and Associated Accessories Needed for Various Cryogenic Coolers — 225

8.0	Introduction		225
8.1	Cryocooler Requirements for Space-Based Communications and Surveillance Imaging		226
	8.1.1 Cooler Requirements for Front-End Components		226
8.2	Cryocooler Requirements for Military Applications		227
	8.2.1 Tactical Coolers for Infrared Missiles		228
8.3	Dilution Refrigeration Systems for Scientific Research		229
8.4	Cryocoolers for Higher Cryogenic Temperatures		229
	8.4.1 Mechanical Refrigerator (MR)		229
	8.4.2 Magnetic Refrigerator Systems (MRS)		230
	8.4.3 TE Coolers		230
		8.4.3.1 Performance Parameters of TE Coolers	231

8.5	Heat Pipe Concept for Higher Cryogenic Temperatures		232
	8.5.1	Performance Limitations Due to Working Fluids in Heat Pipes	234
8.6	Impact of Material Properties on Cooler Performance		237
	8.6.1	Properties of Various Materials Used in Cryogenic Coolers	238
	8.6.2	Thermal Properties of Materials at Cryogenic Temperatures	238
	8.6.3	Electrical Properties of Materials at Cryogenic Temperatures	241
	8.6.4	Mechanical Properties of Materials at Cryogenic Temperatures	246
8.7	Characteristics of Potential Refrigerants		248
	8.7.1	Cooling Capacities of Liquid Cryogens	250
	8.7.2	Cooling with Solid Cryogens	250
8.8	Maintenance Requirements for Various Cooler Accessories		251
	8.8.1	Requirements for Critical Accessories	251
	8.8.2	Cryogenic Insulation Requirements	253
	8.8.3	Impact of Cryogenic Leak in Tubes, Fittings, and Valves	253
	8.8.4	Impact of Thermoacoustic Oscillations on Cryogenic Coolers	254
8.9	Summary		254

Index 257

Chapter 1 | Technological Advancements and Chronological Development History of Cryogenic Technology

1.0 Introduction

This chapter focuses on the chronological development history and technological advancements in cryogenic technology. Cryogenics, a branch of low-temperature physics concerned with the effects of very low temperatures, was first investigated by Michael Faraday, who demonstrated that gases could be liquefied leading to the production of low temperatures of around 173 K. Cryogenic technology gained widespread recognition during the 1960s with emphasis on cryogenic techniques, cooling system installation configurations, and applications. Critical design issues and aspects — including liquefaction of liquid helium, insulated cryogenic containers, improved low-temperature bearings, seal and lubrication techniques, superconductivity technology, magneto hydrodynamic effects, and large-scale industrial applications — were given serious consideration. A comprehensive tutorial treatment was given to fundamental thermometry and accurate measurement of ultralow temperatures for the benefits of cryogenic engineers and of scientists deeply involved in low-temperature research activities.

Studies performed by the author indicate that cryogenic cooling plays an important role in the design and development of cryogenic systems or sensors for commercial, industrial, military, and space systems including missile tracking sensors, electrical machines [1], unmanned aerial vehicles (UAVs), infrared (IR) search and track sensors, satellite tracking systems, pollution monitoring sensors, imaging sensors using focal planer array (FPA) technology, high-resolution imaging sensors, forward-looking IR sensors, multispectral and hyperspectral sensors, magnetic resonance imaging (MRI) and computer tomography (CT) equipment for medical diagnosis and treatment, and high-power laser-based sensors for missile tracking. The studies further indicate that cryogenic technology has potential applications to photonic devices, electrooptic components, millimeter wave and radio frequency (RF) sensors, and optoelectronic devices [2].

1.1 Terms and Phenomena Associated with Cryogenic Systems

Readers are encouraged to familiarize themselves with the various terms and phenomena associated with cryogenic engineering and systems in order to appreciate and comprehend the various operational aspects and benefits of cryogenic technology. The most widely used terms and phenomena associated with the design, development, and operation of cryogenic systems and their critical elements can be summarized as follows:

- Cryogenic liquids
- Cryogens
- Gas liquefaction
- GT-cycle
- JT-expansion valve
- Adiabatic expansion
- Irreversible adiabatic expansion
- Isentropic expansion (based on constant entropy)
- Isenthalpic expansion (based on constant enthalpy)
- Radiated heat transfer
- Cryogenic temperature
- Cooling or heat load capacity
- Cooling efficiency

1.2 Prominent Contributors to Cryogenic Technology

The most prominent contributors to cryogenic technology include five distinguished physicists and chemists, namely, Andrews, C. de la Tour, Faraday, Joule, and Thomson. Between 1820 and 1870, the latter two demonstrated the dependence of the gas's energy on the operating pressure and temperature. Andrews of Ireland performed a series of experiments in 1869 with carbon dioxide and discovered a critical temperature above which the liquid state cannot exist regardless of pressure. Later

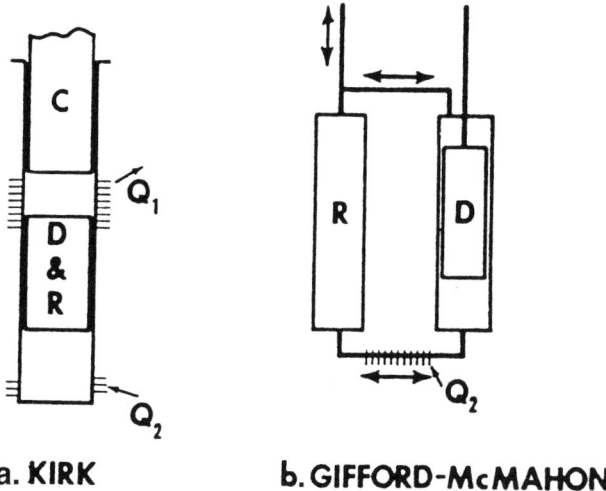

a. KIRK **b. GIFFORD-McMAHON**

SYMBOLS DESCRIPTION:

Q = amount of heat rejection during compression

D = displacer

R = regenerator

C = compressor involving a cylinder and piston

Fig. 1.1 Revised Stirling cycles used by Kirk and Gifford-McMahon (GM).

research scientists studied the properties of permanent gases. A French scientist, Cailletet, liquefied oxygen for the first time in 1877; Polish physicists Wroblewski and Olszewski liquefied oxygen and air in large quantities during the 1880s; an English physicist, Dewar, first liquefied hydrogen gas in 1898; and a Dutch physicist, Kamerlingh-Onnes, first liquefied helium gas in 1908 [3].

While Faraday was busy with experiments to liquefy gases, the Stirling brothers in Scotland developed a sophisticated hot-air engine during the years 1825 to 1840. When this engine reversed and was used as a heat engine, it led to other important cryogenic developments. Scottish scientist Kirk developed an impressive chiller for cold storage and ice making during the early 1860s by adapting the Stirling cycle for refrigeration (Figure 1.1). Right after that, around 1880, more compact and efficient ammonia-condensing systems were developed. Dutch scientists including Kohler and others tried to improve the performance of the Kirk chiller using an efficient air liquefier developed between 1948 and 1954, which provided the refrigeration capability in the 20 K range. The Kirk cycle shown in Figure 1.1 was later modified by Gifford and McMahon (G-M) in 1959 using a remotely located system composed of a compressor, regenerator, and expansion device.

Ice-making machines used by Florida physician Gorrie led to the development of both the expansion engine and the counterflow heat exchanger, which were found to be of significant importance in cryogenics. Machines developed by Gorrie and Kirk [3] were competing with each other until ammonia refrigeration replaced both of them in the 1880s. Siemens in 1857 recognized the possibility of achieving low-temperature capability using the regenerator developed by Stirling and the counterflow heat exchanger invented by Gorrie. After 40 years passed, between 1899 and 1902 scientist Claude was able to liquefy air using an improved version of the engine developed by Gorrie. The 40-year lag was due to the lack of scientific interest in achieving very low temperatures during the mid-19th century. The evolution of modern refrigeration cycles based on the Kirk and Stirling cycles are shown in Figure 1.1. Note that the significant difference in the G-M variation of the Sterling cycle lies in the complete separation of the compression unit from the expansion device.

The simple Linde cycle, based on the Joule-Thomson (J-T) effect, was employed to liquefy air around 1895 by distinguished scientists including Linde, Hampson, Dewar, and Kamerlingh-Onnes. Dewar used the Linde cycle to liquefy hydrogen gas in 1898. Precooling the liquid air lowered the temperature of the hydrogen below the inversion temperature point. It is important to mention that Kamerlingh-Onnes first liquefied helium gas in 1908 using this method. The very low inversion temperature of the helium (32 K) required precooling with liquid hydrogen.

According to the published literature, in as early as 1840 John Gorrie built the first refrigeration machine to achieve its cooling power from the expansion of the compressed gas in an engine. Claude further improved this refrigeration cycle, and it was used successfully to demonstrate air liquefaction capability in 1902. Claude designed hydrogen expansion engines using the isenthalpic expansion technique during the 1920s to produce cryogenic temperatures below the liquid air level. It is interesting to mention that the first use of helium by Kapitza as a working fluid

(1934) was the most significant contribution to the art of the J-T heat exchanger to further cool the stream of compressed helium already cooled by the expanded helium from the J-T engine. In 1946, Collins demonstrated the helium liquefier with isentropic expansion of helium at two temperature levels, thereby eliminating the need for precoolants in the liquefaction of the helium. Between 1927 and 1933, Simon demonstrated the liquefaction of small quantities of helium effectively using oxygen at a 300 atm pressure and temperature between 253 and 233 K. Simon recognized that a high pressure charge of helium, cooled by hydrogen, almost fills the container with liquid helium when the operating pressure falls to one atmosphere or 14.7 p.s.i. or 760 mm of mercury.

1.3 Critical Aspects and Issues Involved in Cryogenics

Critical aspects and issues need to be addressed prior to describing the potential applications of cryogenic technology. Cryogenic technology [4] involves the following critical aspects and issues:

- Theory of cryogenic electrons and phases
- Low-temperature physic theory
- Cryoelectronics, including superconducting and nonsuperconducting circuits
- Application of cryogenic technology to space sensors
- Materials and fluid properties and their influence on the design of low-temperature, cryogenically cooled equipment
- Cryogenic wind tunnels for aerospace applications
- Cryogenics for medicine and biological applications
- Technological aspects of liquid helium, hydrogen, and neon gas
- Cryogenic applications in military, space, and industrial systems
- Cryogen separation and gas separation technique
- Superconducting magnets for MRI, heavy equipment lifting, and levitation trains
- Cryogenically cooled devices and sensors for scientific, industrial, commercial, and medical diagnostic applications

The key words most widely used in cryogenics include Stirling cycle, Gifford-McMahon cycle, J-T closed cycle, cryogen, cooling capacity, Dewas, maintenance Breelife, and heat exchanges. Readers who wish to expand their horizon in the field of cryogenics will benefit from becoming familiar with these key words.

1.4 Benefits from Integration of Cryogenic Technology

Cryogenic technology, when integrated in electronic devices or sensors, offers significant improvement in critical performance parameters. However, affordability, availability, and reliability issues must be given serious consideration prior to integrating this technology in sensors or systems for specific applications.

1.4.1 AFFORDABILITY

In any system, the performance improvement due to the integration of cryogenic technology must be affordable. Affordability encompasses not only acquisition and life cycle costs but also includes the costs associated with installation, operation, and maintenance. Studies performed by the author indicate that the weight, volume, and electrical power consumption of a cryogenically cooled system, including the cryogenic and cooling elements, will be less than one-third of a conventional system capable of performing the same functions. Integration of cryogenic technology in some systems can realize reduced cost and complexity, if the cryogenically cooled system is properly designed and installed. Cryogenically cooled circuit fabrication uses processes similar to those commonly used in the production of gallium arsenide (GaAs) integrated circuits (ICs), but it requires significantly fewer processing steps, leading to considerable savings in the fabrication process. Note that pulse tube cryogenic coolers with projected mean time between failure (MTBF) exceeding 10,000 hours are best suited for airborne and space system applications, where reliability and improved system performance are of critical importance. Higher cryogenic cooler reliability coupled with lower operating temperature should significantly improve the system's reliability, leading to a considerable reduction in both the maintenance and life cycle costs.

1.4.2 AVAILABILITY

The enhanced reliability over conventional technologies due to the integration of cryogenic technology coupled with modular line replaceable cooler/electronics units should improve availability. In addition, superconducting screens or shutters can contribute significantly to increased subsystem hardiness and operational availability. Thus, improved availability should result in the wide acceptance of cryogenic technology in commercial, industrial, military, and space-based systems.

1.4.3 PERFORMANCE IMPROVEMENT IN VARIOUS COMPONENTS AND SYSTEMS

Significant performance improvements have been observed in cryogenically cooled electronic circuits, microwave and mm-wave components, electrooptical devices, and IR sensors deployed in commercial, industrial, and military applications. Improving the performance of these devices and sensors will not be possible without the integration of cryogenic technology. Studies performed by the author reveal that cryogenic technology offers several advantages in the telecommunications systems, such as wide communication bandwidth, high data rates, high-speed response, low component or systems losses, higher reliability, low cross-talk in telecommunication systems, reduced weight and size, and minimum power consumption. However, tradeoff studies must be performed before implementing such technology in a device or system to satisfy the cost-effectiveness criteria for a specific application. Many electronic and digital technologies — such as complementary metal-oxide semiconductor (CMOS), GaAs, metal semiconder field effect transistor (MESFET), high electron mobility transistor (HEMT), optoelectronic, electrooptic, and even ordinary metals including aluminum and copper — have demonstrated significant performance improvement at lower operating temperatures. However, with the use of certain superconducting materials and an optimum cryogenic temperature, further performance improvement can be realized at the component [1] or system [2] level. When the advantages of superconducting electronics are complemented by the advantages of other cryogenic technologies, the result is a hybrid system whose overall performance improvement can justify the additional cost and complexity of cryogenic cooling.

1.5 Early Applications of Cryogenic Technology

This section briefly discusses the potential applications of cryogenic technology in commercial, industrial, medical, space, and military fields. The use of cryogenic technology for the production of gases for industrial and commercial use, for wind tunnel tests, and for other industrial applications is described.

1.5.1 CRYOGENIC TECHNOLOGY FOR THE PRODUCTION OF GASES

UNOCAL Corp. of Utah first started the production of gases for industrial and commercial applications. The company has a gas treatment plant and a helium (He) purification and liquefaction plant that have been in operation since 1991 with a liquefied helium shipping capacity of two to three 11,000-gallon truckloads per week. The overall U.S. production capacity in 1991 was more than 3500 tons per day of oxygen and 3000 tons per day of nitrogen. By the end of the year 2000, the Bayport Company of Texas had the largest air gas plants in the world with total capacity exceeding 10,000 tons per day of oxygen and nitrogen.

1.5.2 CRYOGENIC TECHNOLOGY FOR THE PRODUCTION OF INERT GASES

The production of inert gases accelerated right after World War II due to the heavy demand for scientific research and military applications. Air Products and Chemicals of Hometown, Pennsylvania, installed a nitrogen trifluoride (NF_3) plant capable of producing more than 100,000 pounds per year. The company uses its own technology and a unique purification process to achieve a high purity product with low impurity contents. This inert gas of high purity is best suited for the electronics industry and is widely used as a fluorine source in high-energy chemical lasers, in situ cleaning processes for chemical vapor deposition equipment, and semiconductor circuit etching techniques.

1.5.3 CRYOGENIC TECHNOLOGY FOR AEROSPACE APPLICATION

Cryogenic technology [5] is very critical for wind tunnel testing applications. The wind tunnel designs in 1991 were limited in the extent to which they could model the combination of size and speed of the modern supersonic aircraft and missiles. High-performance wind tunnels require rapid movements of nitrogen gas around the wind tunnel's aerodynamic circuit. To meet the stringent test requirements, more than 900 tons per hour of liquid nitrogen will be sprayed into the tunnel. Nitrogen supplies will initially come from the 3000-ton storage tanks, which can be upgraded to meet higher gas demand.

1.5.4 CRYOGENIC LIQUID LEVEL CONTROLLER (LLC)

The cryogenic liquid level controller (LLC) plays a critical role in the operation and maintenance of cryogenic storage tanks, because it offers an instant indication of the cryogenic liquid level present in the storage tank. This device provides an accurate cryogenic liquid level better than 0.1% and yields reliable continuous level indication of virtually all cryogenic liquids by measuring the sensor capacitance, which is directly related to the cryogenic liquid level. The LLC provides continuous level measurements over the entire range of the sensor. These LLC sensors are available in lengths of up to 16 feet with a variety of standard and custom mounting configurations. The important sensor features include a digital display, light emitting diode (LED)-based alarm indicators and alarm relays, analog recorder output, and nonvolatile read-only memory.

1.5.5 CRYOGENIC LINE REGULATORS

Some industrial applications require high-purity, single-stage line regulators with superior leak integrity without contamination from nonmetallic lines or seals. Leak integrity is provided by a convoluted stainless steel diaphragm designed to achieve accurate as well as stable delivery pressure. Such a line regulator is capable of withstanding internal vacuum purging with high reliability. It is important to mention that ultrasonic cleaning permits high-purity gas handling capability without costly precleaning procedures, and high flow capacity allows accurate pressure control for multi-instrument applications where several parameters need to be monitored at the same time. This particular line regulator has an optical bonnet vent adapter capable of allowing the hazard gases to be vented in the event of diaphragm failure and a provision for a bracket or external panel mounting.

1.6 Gas Separation Process Using Cryogenic Technology

The gas separation process is of critical importance for some commercial and industrial applications. Significant improvements have been made in air separation processes involving cryogenic and alternate technologies. The double-column, dual-reboiler cycle technology developed by Air Products and Chemicals Company in Allentown, Pennsylvania, saves much of the energy wasted in conventional cryogenic production of medium-pressure (6- to 10-bar atmosphere) gaseous nitrogen from air. Much of the energy associated with an oxygen-rich stream is lost and only a part of the stream is turbo-expanded for refrigeration. In a modified system using pressure swing absorption (PSA) techniques [5], the portion not turbo-expanded is recycled to air feed stream, so that the pressure is not wasted. This recycling partially condenses the overhead nitrogen in the first of the two condensers and has its own nitrogen concentration stepped up to near that of the air by the distillation trays mounted above the condensers. Integration of the PSA technology with advanced carbon molecule sieve technology offers cost-effective operation yielding 99.9% pure nitrogen at production rates in the range of 500 to 40,000 specific cubic feet per hour.

The gas separation system developed by Union Carbide Corporation involving vacuum pressure swing absorption (VPSA) technology and propulsion zero lite technology produces nitrogen gas up to 99.95% pure and oxygen up to 99.5% pure. The company claims that its system's capital costs are 25% to 35% less than a PSA-based gas separation system.

1.7 Industrial Applications of Cryogenic Fluid Technology

Several industrial applications of cryogenic fluids have recorded [6] significant progress in the 1960s and 1970s. New developments have used liquid hydrogen, neon, and nitrogen as refrigerants because of maximum economical and safety reasons. According to published data, most popular refrigerants include liquid neon (27.1 K), liquid nitrogen (77.3 K), liquid helium (4.2 K), and liquid hydrogen (20.3 K). Note that the values shown in parentheses indicate the critical temperatures of the refrigerants [7].

1.7.1 LIQUID NEON

Liquid neon is widely used for commercial and industrial applications and is considered the most ideal substitute for liquid hydrogen due to its chemical inertness. As technological advancements in cryogenic science unfold and as the size of the tonnage air separation plants increases, neon availability will start to exceed demand and reduce production prices, which will lead to significant popularity for various applications.

1.7.2 LIQUID HYDROGEN

Liquid hydrogen has been widely used in applied cryogenics because of its minimum cost when produced in large volumes. However, its adverse chemical effects and stringent handling requirements largely offset this potential. It is widely used as a propellant in rockets, boosters, missiles, and space vehicles.

1.7.3 LIQUID NITROGEN (LIN)

Liquid nitrogen is a fairly inert gas medium and has unique properties that make it a most ideal space simulation chamber and cold drawing of stainless steel used in specific industrial and scientific research applications. The following three distinct developments promise a great demand for LIN, where exotic and large-volume commercial applications are involved:

- Deflating of molded rubber parts requires a process known as CRYO-TRIM.
- Refrigeration for trucks, trailers, and railroad carts for in-transit preservation of fruits, vegetables, meats, and other perishable food items requires a process known as CRYO-GUARD.
- Freezing of baked goods, shrimp, TV dinners, meats, soups, and so on requires a process known as CRYO-QUICK [7] in the food processing industry.

As far as the heat absorbing capacity is concerned, liquid hydrogen absorbs about 192 BTU/lb, while liquid nitrogen absorbs around 86 BTU/lb. Liquid nitrogen is widely used in food processing and food transportation because of minimum production costs. It is important to mention that two unique characteristics of LIN — namely, vaporization from liquid to gas absorbing heat energy of 86 BTU/lb of LIN and an expansion ratio as high as 697 — provide thorough penetration essential for gas temperature measurement. This means that process temperature control is possible with small volume and with minimum cost. The exit nitrogen temperature must be raised to as high a temperature as possible for maximum utilization of all available refrigeration, which will lead to a most economical operation.

1.8 Heat Capacity of Commercial Refrigerants

LIN is an economical cryorefrigerant, which offers more than 40 times more refrigerating capacity per unit volume than liquid helium and more than 3 times that of liquid hydrogen. It is compact inert gas and is less expensive than helium when it meets the refrigerant cost requirement of less than $1.50 per liter. Helium gas is another critical refrigerant, which costs about $35/1000 cu.ft. Helium (He) can be used as an inert gas shield for arc welding, for growing silicon and germanium crystals by the semiconductor industry, for production of titanium and zirconium as core media in nuclear reactors, for wind tunnel experiments needed in the design and development of supersonic fighter aircraft and missiles, and for pressurizing liquid fuel to be used by rockets, boosters, and missiles. Its two most important isotopes include tritium, an important ingredient used in the development of hydrogen bombs, and deuterium oxide, commercially known as heavy water widely used as a moderator in nuclear reactors. Important properties of widely used cryogenic liquids are summarized in Table 1.1.

Table 1.1 Properties of Cryogenic Liquids Widely Used in Commercial and Industrial Applications.

Cryogen type	Boiling point (K)	Molecular weight	Latent heat (BTU/lb/mole)	Sp. heat (BTU/lb/0F)	Heat energy (BTU/lb)
Hydrogen	20.3	2.02	389	6.89	385
Nitrogen	77.3	28.01	2405	6.98	86
Neon	27.1	2.18	748	4.98	—
Helium	4.2	4.00	—	—	—

1.9 Cryogenic Requirements for the Frozen Food Industry

Cryogenic requirements for the preservation of food items depend on the type of food items and whether they are cooked or raw before freezing. The cooking time and temperature strictly depend on the type of food, the quantity and quality of food to be frozen, in-transition duration, storage time, general public health considerations, and Federal Drug Administration (FDA) guidelines [7]. The quality of a frozen food product is contingent on the natural composition of the food item and the processing changes that occur in various food items such as red meat, poultry, fish, and vegetables after the food is prepared and prior to freezing [7].

Freezing time in a blast freezer varies from 40 minutes to 2.5 hours depending on the food item and its weight. Food taste and its judgment depend on appearance, texture, flavor, and color during preparation. Cooking and handling losses prior to blast freezing vary from 2% to 8%, approximately, depending on the number of meals prepared and processes involved. When considering a cook-freeze program, the following steps must be taken into account to preserve food taste and quality:

- Preliminary estimates must be prepared to make sure that the cost needed to freeze the food per meal is within the acceptable limits set by higher management.
- An accurate cost and accounting system must be developed and implemented.
- Low-cost food items requiring high-cost intensive labor must be given serious consideration to meet the cost guidelines established by management.
- High-cost menu items such as steaks, because of the seasonable availability of their ingredients, must be given top priority to meet customer requirements.

1.9.1 COLD STORAGE REQUIREMENTS

Cold storage requirements depend on the type of item to be stored, safe storage duration, the type of refrigerant used to meet cost-effective criteria, and health guidelines established by the FDA. A cost-storage (CS) system employing off-the-shelf equipment to extract heat from a storage medium will be found most cost-effective, when operating during nighttime hours. The CS medium is then used to absorb the heating space during the normal daytime hours, using only the pump and fan in the process. The CS is categorized based on the type of storage medium used such as an ice-based or water-based medium and the mode of operation depending on the size of storage capacity needed [8]. A typical hourly load profile for a given storage system under full storage operating mode is shown in Figure 1.2.

1.10 Cryogenic Requirements for Medical Applications

Cryogenic technology plays a key role in real-time, high-resolution imaging systems used for medical diagnosis and treatment of a host of illnesses. The performance of magnetic resonance imaging (MRI) equipment widely used in medical clinics and hospitals to diagnose various diseases is contingent on a cryogenic temperature as low as 4.2 K. An MRI system is composed of a large superconducting magnet that is cryogenically cooled by a closed-cycle refrigeration (CCR) system at a temperature of around 4 K. The cooling capacity of the CCR system depends on several operating parameters, resolution capability, and the equipment reliability requirement. The CCR deployed initially in MRI equipment consists of a two-stage Gifford-McMahon (G-M) refrigerator capable of operating at two distinct cryogenic temperatures, namely, 77 K and 20 K, to cool various MRI system components. This cryogenic system provides liquid nitrogen cooling at 77 K for the magnet and at 20 K cryogenic cooling for thermal shielding to reduce the liquid helium evaporation with a helium refilling time of several months. However, the latest three-stage CCR system incorporating a two-stage G-M refrigerator and a J-T expansion valve can recondense helium fully with no need to refill helium. This particular refrigerator system offers the cryogenic performance level summarized in Table 1.2.

Table 1.2 Cryogenic Power Requirements of a Three-Stage CCR System at Various Cryogenic Temperatures.

Cryogenic temperature (K)	Cryogenic power requirement (W)
16	3.50
4.5	0.65

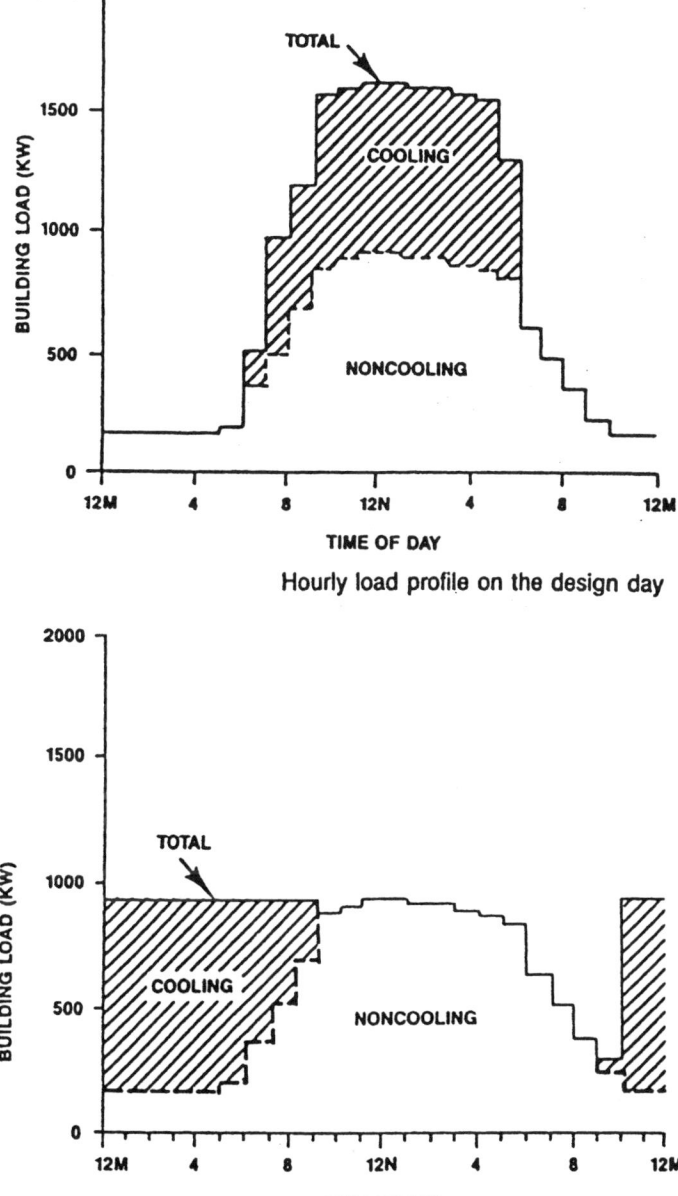

Fig. 1.2 Design day load profile under the full storage operating mode.

The input electrical power was about 4500 watts, which is what was needed to maintain required cryogenic temperatures and cryogenic power or cooling power. This system deployed a rotary vane compressor unit with electrical power consumption between 10 and 12 kW. Excessive voltage fluctuations, use of ferromagnetic materials, and structural vibrations must be avoided to achieve perfect homogeneity and improved MRI image quality by maintaining an optimum cryogenic temperature, regardless of operating environments.

1.10.1 CRYOGENIC SYSTEM REQUIREMENTS FOR HIGH-RESOLUTION MRI

Practically all MRI equipment requires cryogenic cooling well below 5 K to achieve high-resolution imaging for reliable and most acceptable diagnosis results. Current cryogenic refrigeration systems use evaporated liquid helium that is cooled in a batch process involving small superconducting devices. This approach offers improved performance with minimum cost, particularly for systems with large heat loads and long continuous operations. A closed-cycle mechanical refrigerator operating at a cryogenic temperature of 4.5 K provides five times more refrigeration capability or cooling power than a liquid helium vaporization of the same size and same power rating. Furthermore, such a refrigerator system does not require recover and repurification equipment, thereby leading to a most cost-effective cryogenic operation. Cost-effective MRI equipment requires the following design features:

- High reliability under continuous operation requiring minimum maintenance
- Select design configuration that occupies minimum floor space
- Simple operating procedures to minimize the cooldown time
- A system with high cryogenic efficiency to minimize capital investment and operating costs
- A cryocooler design with minimum noise and vibration to provide maximum comfort for the patients

1.11 Industrial Applications of Cryogenic Technology

Cryogenic technology is being used for commercial and industrial applications such as cryopumping, nuclear radiation testing, cold storage for food items, space research studies, complex ice-making machines with ultra-high capacities, and other scientific investigations. These applications may require cooling systems that have refrigeration capacities ranging from 100 to 4000 watts and liquefaction capabilities of 100 to 250 liters per hour depending on the application. Potential applications for widely used commercial and industrial systems will be briefly described with emphasis on cooling capacity and operating costs.

1.11.1 CRYOPUMPING

Studies performed by the author on commercial and industrial applications indicate that large closed-cycle refrigeration (CCR) systems are widely deployed in cryopumping applications to provide the most cost-effective operation in the long run. Cryopumping applications deal with thermal-vacuum space chamber facilities. Several refrigerator systems are employed with low-density wind tunnels in gas dynamic tests, which are most critical in the conceptual design phase of supersonic jet fighters and missiles. Most of the cryopumping refrigerators operate at a temperature of around 20 K, with cooling capacities ranging from 400 to 4000 W depending on the specific application.

1.11.2 NUCLEAR RADIATION TESTING

Nuclear radiation level testing is required to monitor the radiation levels of nuclear power plants and nuclear weapon systems under various operating conditions. Radiation tests are needed to assess the nuclear leakage or nuclear irradiation damage caused by high-energy gamma and neutron radiations. Reliable tests are performed randomly at temperatures as low as possible to achieve better testing accuracy [9]. The cooling of cryogenic sensors and experiments are done at remote location to avoid exposing nearby personnel. For this particular application, refrigeration systems use extremely pure helium gas for high accuracy. Gamma radiation in these experiments sometimes can cause a heat load of high intensity. CCRs using pure helium gas with operating capacities of around 1000 W at 4 K cryogenic temperature, and 1800 W at 18 K operations, are widely deployed to conduct such tests at nuclear reactor facilities. Refrigeration systems operating at a cryogenic temperature of 4 K generally require liquefiers with helium capacities ranging from 50 to 150 liters per hour. Demand for this amount of refrigerant justifies operating a plant for com-

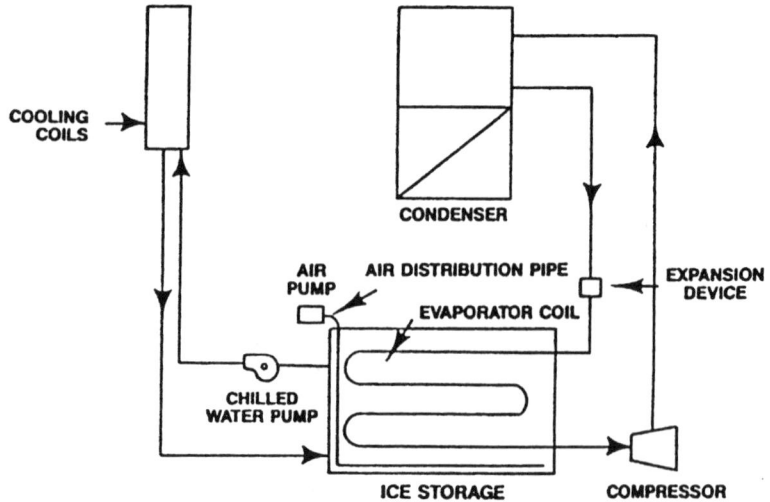

Fig. 1.3 Critical elements of a basic ice storage system.

mercial production of liquid helium and distribution of the same within the nuclear research complex facility or test site to avoid logistic and storage problems.

1.11.3 ICE-MAKING MACHINES AND ICE STORAGE SYSTEMS

Only ice-making systems can meet the heavy demand for ice by consumers, commercial firms, and grocery stores. Static ice-making systems are available in various capacities, typically ranging from 50 to 1500 tons per hour depending on the consumption and weather environments. All packaged units can be connected to an existing or future chilled water system. In the basic static ice storage system, as illustrated in Figure 1.3, a water/glycol solution is used as a circulating working fluid to lower the operating costs. However, a modular ice storage system, as shown in Figure 1.4, using a brine circulating fluid offers high production efficiency because of higher ice-to-water ratio.

Combining the ice-making and storage functions in a single system configuration offers maximum economy and a compact system design. The sprayed freezing coil ice-making storage system shown in Figure 1.5 offers both the ice-making and storage capabilities with minimum cost, supervision, and complexity. Critical system elements such as cooling coil, compressor unit, condenser assembly, freezing coil, spray circuit, and

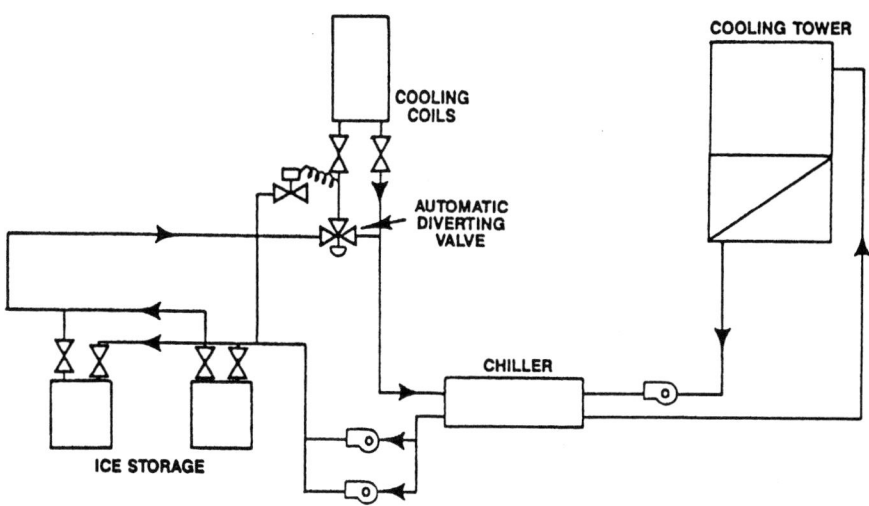

Fig. 1.4 Critical elements of a modular ice storage system using brine.

ice storage are clearly shown in Figure 1.5. This particular system [9] uses spraying water from a storage bin over sprayed freezing coils, which offers higher cooling efficiency. Because these coils are extended to the storage bin, more bin volume can be filled with ice that will lead to a smaller freezing surface.

When the solution used is cooled down to below its freezing point temperature, small ice crystals are formed producing ice slurry, which is the most salable product of its kind during the summer. A dynamic ice slurry production system [9] — along with system elements including cooling coils, an ice storage unit, a slurry circulating pump, and a slurry ice generator — is depicted in Figure 1.6. High slurry production volumes must be made available in summer to meet the higher demand by school children, which will bring additional revenues.

1.11.4 CHILLED WATER STORAGE (CWS) SYSTEM

Operation of the CWS system shown in Figure 1.7 depends on effective utilization of the heat capacity of water. The CWS system employs a conventional chilled water-cooling system augmented by a chilled water storage tank, as illustrated in the figure. The critical elements of a CWS system include a chiller, chilled water pumps (primary and secondary),

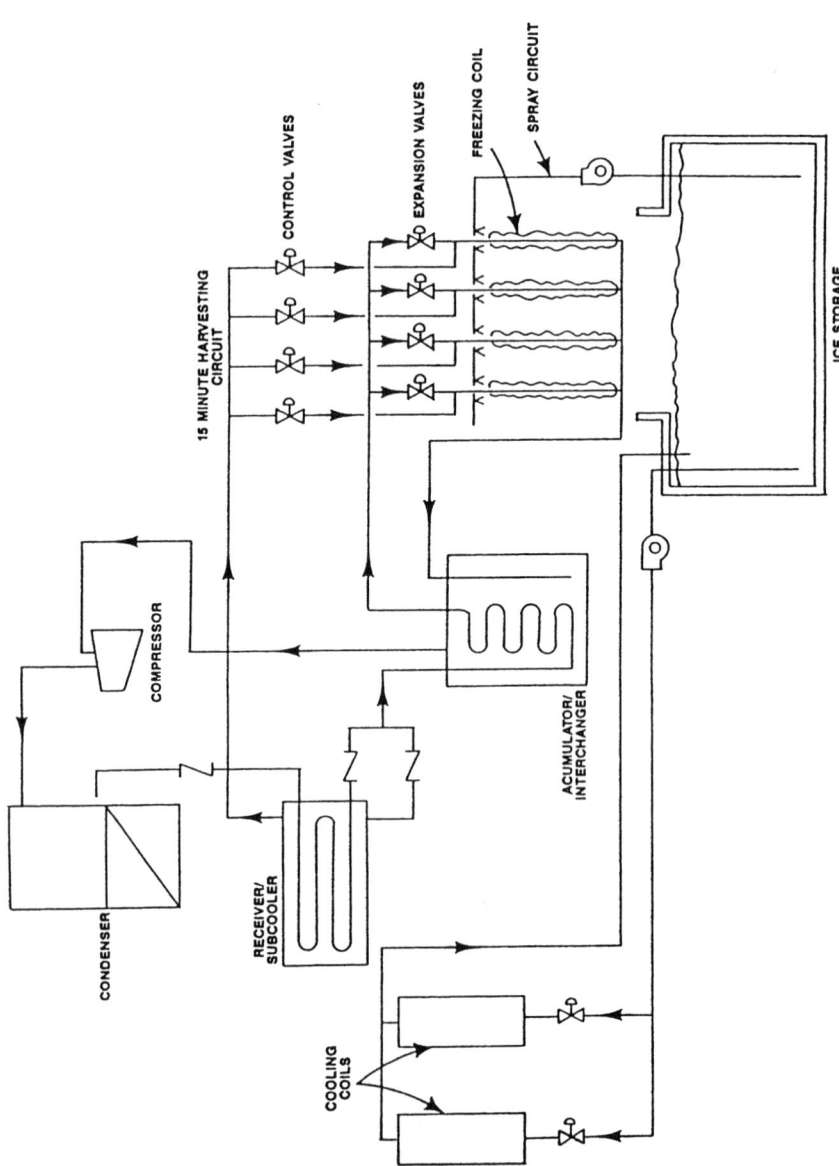

Fig. 1.5 Elements of a sprayed freezing coil ice-making storage system.

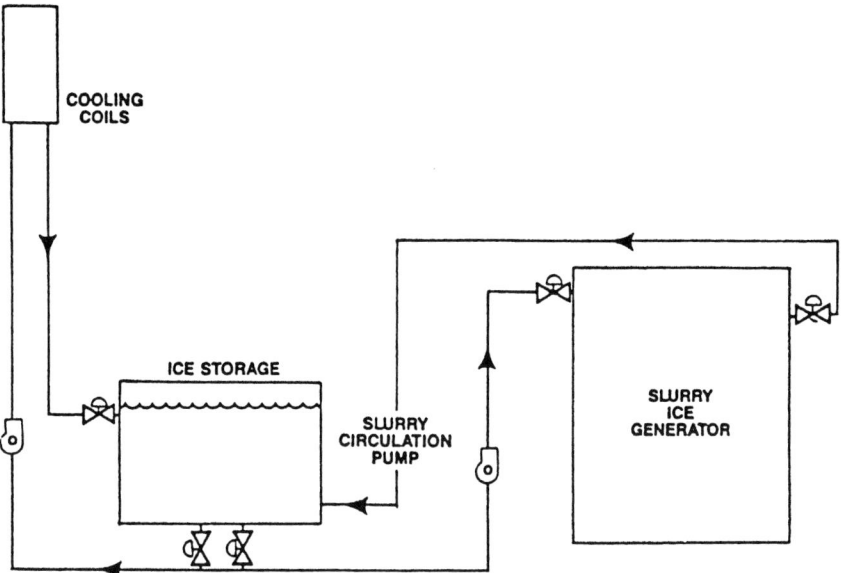

Fig. 1.6 Block diagram showing the critical elements of a dynamic ice slurry system.

cooling coils, a cooling tower with reasonable height, a chilled water storage tank, and a condenser water pump [9]. It is important to point out that the chilled water storage tank must be larger than the ice storage tank for the same cooling capacity. In some applications, the storage tanks used for chilled water storage also are used to store hot water produced by using energy during off-peak periods, thereby yielding additional economic benefits. Note that a storage system must provide enough storage to meet full on-peak season cooling requirements. Cooling demand varies according to the time of the day and the environmental temperature at that time. A typical load profile [9] for the full-storage operating mode is shown in Figure 1.2.

1.12 Summary

This chapter briefly described the technological advances and chronological development history of cryogenic technology and engineering. Applications of cryogenic technology in earlier commercial and industrial systems were discussed. Critical design issues and the operational aspects

22 Cryogenic Technology and Applications

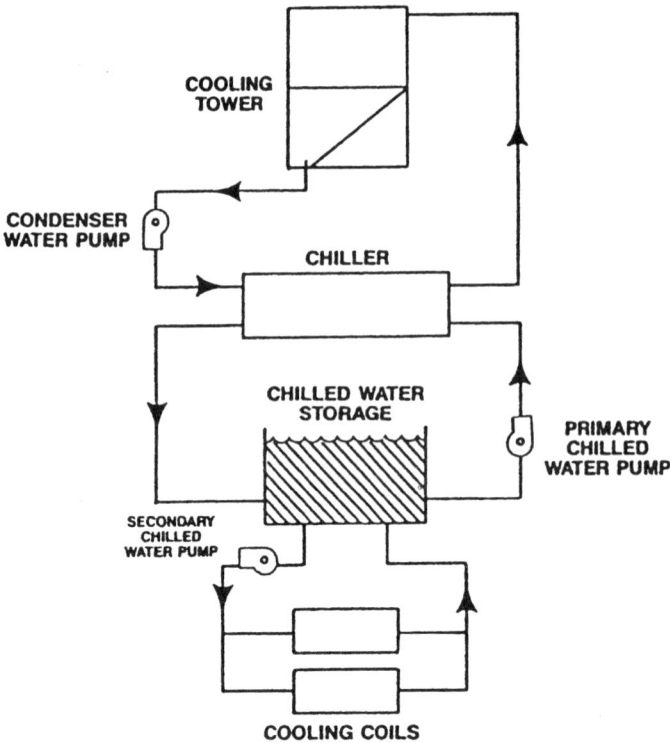

Fig. 1.7 Elements of a chilled water storage system.

of cryogenic systems were briefly discussed, with an emphasis on cost and reliability. Significant performance improvements in devices, components, and systems due to the integration of cryogenic technology were highlighted, with major emphasis on reliability, affordability, and availability. Cryogenic technology for the production of inert gases for industrial, commercial, and military applications was briefly discussed, and the design issues necessary to provide safe operations were identified. Important properties of selected cryogenic liquids widely used are summarized. Applications of cryogenic technology in cryopumping, cold storage, ice making, medical diagnosis, nuclear radiation assessment, and production of chilled water and slurry were explored in greater detail.

References

1. A. R. Jha. *Superconductor technology: Applications to microwave, electro-optics, electrical machines, and propulsion systems.* New York: John Wiley & Sons, 1998, pp. 242, 245.
2. A. R Jha. *Infrared technology: Applications to electro-optics, photonic devices, and IR sensors.* New York: John Wiley & Sons, 2000, pp. 216, 218, 220.
3. S. C. Collins. *Early history and developments of cryogenics.* Cambridge, MA: Arthur D. Little, 1965, pp. 3, 4, 8.
4. H. Weinstock. *Cryogenic technology.* Cambridge, MA: Boston Technical, 1969, pp. 76, 79.
5. J. R. Campbell and Associates. *Cryogenic information report.* 5 Militia Drive, Lexington, MA 02173 (Fax: 617-863-9411).
6. R. E. Hill. *Large scale industrial applications of cryogenics.* Cleveland, OH: Burdett Oxygen Company, 1970, pp. 121, 124.
7. E. L. Harder. *Blast freezing techniques and recipes.* Boston: CBI Publishing, 1938, pp. 8–9.
8. Editor. *Commercial cold storage design guidelines, by Electronic Power Research Institute, EM-3981.* New York: Harper & Row, 1987, pp. 6, 10.
9. C. Guoban and T. Flynn. *Cryogenics and refrigeration proceedings of international conference.* May 22–23. New York: Pergamon Press, 1989, pp. 107–109.

Chapter 2 | Effects of Heat Flows on Heat Exchanger Performance and Cooler Efficiency

2.0 Introduction

The thermodynamic aspects and heat transfer issues associated with transient and turbulent heat flows are of critical importance, because they have a significant impact on the heat exchanger performance and cryogenic cooler efficiency. This chapter briefly describes the impact of thermal processes; phenomenological thermomechanics; thermomechanical effects; mechanical laws of conservation including mass, energy, momentum, and moment of momentum; and energy balance on cryogenic systems. Internal energies of cryogenic gases and their impact on refrigeration efficiency are discussed. It is important to mention that the method of phenomenological thermodynamics allows the internal energy of atomic substances to be broken into several distinct components. These energy components include thermal energy due to random translational motion, repulsive energy due to mutual repulsion during collisions, attractive energy due to mutual attraction during translational motion, vibrational energy due to elastic longitudinal and transverse isothermal waves, spin energy due to an atom's mechanical rotation about its axis, and surface energy at the interface of the finite macrosystem. Note that in phenomenological thermodynamics, the energy balance equation representing contributions from all of the energy components specified is considered to be the first law of thermodynamics. This particular equation can be expressed in terms gas constant, operating temperature and pressure, the compressibility factor, refrigerant density, Boyle's temperature, and the critical temperature of the refrigerant involved.

2.1 Early Developments in Cryogenic Cooler Technology

The development of refrigeration started in the middle of the 17th century with a scientific experiment [1] conducted by the Englishman Robert Boyle (1627–1691) to demonstrate the reduction of the boiling temperature of water under reduced pressure. Dr. William Cullen at Glasgow University observed in 1755 that an isolated water container showed a drop in temperature during evaporation. Thomas Masters of England in 1844 demonstrated the ice cream maker performance using a mixer of ice and salt to lower the brine temperature close to the freezing point. American scientist Charles E. Monroe of Cambridge, Massachusetts, demonstrated in 1871 the food cooler operational capability based on the evaporation of water from the porous lining of a refrigerator. The concept of a natural refrigerator based on the cutting and storage of ice was first described by Frederic Tudor (1783–1864). Mathematics professor Sir John Leslie (1766–1832) at Edinburgh University in Scotland used sulfuric acid to absorb water vapor from a water container, thereby producing a vacuum in the closed container. The vacuum in turn caused the saturation temperature of the water producing the vapor to be low enough for the formation of the ice. Mass scale ice blocks were made from 1810 to 1881 for commercial applications and for human consumption [1].

Ferdinand P. E. Carre demonstrated in 1824 that ammonia could reach much lower temperatures than water when boiled at the same pressure. In 1930s, the Crosley system based on Carre's cycle was sold widely in the United States for domestic applications. Note that the gas refrigerator operates on the same cycle in a continuous fashion. In 1755, M. Hell observed that air leaving a pressurized air line cooled when it escapes from the line. A refrigeration machine was designed on Howell's observation and a U.S. patent was granted to Dr. John Gorrie (1803–1855) for the first machine to operate successfully on the air compression-expansion cycle.

Jacob Perkins (1766–1842), an American scientist living in London, demonstrated that cryogenic working fluids rather than air could operate more efficiently if they could be condensed easily after compression. Perkins designed the first practical vapor-compression machine, which was patented in 1934. David Boyle of Chicago successfully demonstrated the use of ammonia in such as machine between 1869 and 1873. Carl P. G. Linde (1842–1934) of Munich, Germany, developed a refrigeration machine using an advanced cycle with much better mechanical design features. The machine was developed to an experimental stage in 1873 and went into mass-scale production for commercial use in 1875.

2.2 Impact of Thermodynamic Aspects on Cryogenic Coolers

2.2.1 INTRODUCTION

A clear understanding of the aspects of thermodynamics is absolutely necessary to thoroughly comprehend cryogenic cooler technology. Heat transfer depends on critical parameters of the heat balance equation and fluid flow types such as laminar flow through a tube or channel with specific cross section or turbulent flow. The first and second laws of thermodynamics are of paramount importance. The fundamental thermodynamic law states that when mechanical work is transformed into heat or heat into work, the amount of work is always equivalent to the quantity of the heat (Q). Reversible and irreversible processes in the regenerator play a key role in the design of cryogenic systems. The energy balance equations in a heat transfer system lead to a characteristic length and a characteristic time. In the case of an ideal heat exchange between the gas and the matrix, the equations imply solutions in which the temperature profile moves through the regenerator as a heat wave. The cooling power of a cryogenic cooler is determined by the work done by the compressor and the entropy produced by the irreversible processes in various components of the system. Thermodynamic aspects of pulse tube–based systems as well as most popular cryocoolers including Stirling coolers and Gifford-McMahon coolers will be discussed later in the book.

2.2.2 SYMBOLS AND FORMULAS WIDELY USED IN THERMAL ANALYSIS

The most common symbols used in thermodynamics are summarized in Table 2.1. Partial derivatives of the first order for the eight fundamental thermodynamic variables shown in Table 2.1 may be obtained when a specific variable is kept constant. Thermodynamic formulas [4] in terms of partial derivatives are shown in Table 2.2. These formulas are of significant importance in solving classical thermodynamic equations. Useful conversion factors are summarized in Table 2.3. Important properties of

Table 2.1 Thermodynamic Variables and Their Symbols and Units.

Thermodynamic parameter	Symbol	Units
Pressure	P or p	p.s.i
Temperature	T	K
Volume	V	cm^3
Entropy	S	BTU/K
Internal energy	U or E	Ergs or calorie
Total heat or enthalpy	H or Q	Calorie, BTU
Helmholtz free-energy	A (= U − TS)	Ergs
Gibbs' free-energy	G (= H − TS)	Ergs

cryogenic gases are shown in Table 2.4, which will be found most useful in solving thermodynamic problems.

2.3 Types of Heat Flow

Heat transfer depends on the operating parameters, flow pipe cross-sectional geometry (circular or rectangular or square), and fluid flow types such as laminar, turbulent, or transient. Heat transfer properties for a linear flow are quite different from those for a turbulent flow or transient flow. Under the turbulent flow environments, the flow will be treated as a nonhomogenous flow, and, therefore, nonlinear flow equations will be involved in the thermal analysis of heat exchangers. One- and two-dimensional heat flow problems will be addressed later in this chapter. Regardless of whether it is a water flow or heat flow, one has to deal mainly with two types of flow: linear flow and turbulent flow. The rotation of the tube carrying the fluid does not affect the laminar flow resistance once the established cooling flow exists. However, under the turbulent flow conditions, flow resistance undergoes radical change. A numerical model is required to study the unique characteristics and features of coolant flow and heat transfer within the superfluid refrigerant medium such as helium. Note that the rotation of the tube-containing refrigerant does not affect the laminar flow resistance. The conservation equations for dimensional stream function, vorticity transport, axial velocity, and temperature can be manipulated in terms of the rotational Rayleigh

Table 2.2 Thermodynamic Formulas.

		Constant		
Differential	**T**	**P**	**V**	**S**
T	0	1	$\left(\dfrac{\partial V}{\partial p}\right)_T$	$-\left(\dfrac{\partial V}{\partial T}\right)_P$
P	-1	0	$-\left(\dfrac{\partial V}{\partial T}\right)_P$	$-\dfrac{nC_p}{T}$
V	$-\left(\dfrac{\partial V}{\partial T}\right)_T$	$\left(\dfrac{\partial V}{\partial T}\right)_b$	0	$\left(-\dfrac{1}{T}\right)\left[nC_p\left(\dfrac{\partial V}{\partial T}\right)_T + T\left(\dfrac{\partial V}{\partial T}\right)_p^s\right]$
S	$\left(\dfrac{\partial V}{\partial T}\right)_p$	$\dfrac{nC_p}{T}$	$\left(\dfrac{1}{T}\right)\left[nC_p\left(\dfrac{\partial V}{\partial p}\right)_T + T\left(\dfrac{\partial V}{\partial T}\right)_p^s\right]$	0
U	$T\left(\dfrac{\partial V}{\partial T}\right)_p + p\left(\dfrac{\partial V}{\partial p}\right)_T$	$nC_p - p\left(\dfrac{\partial V}{\partial T}\right)_p$	$nC_p\left(\dfrac{\partial V}{\partial p}\right)_T + T\left(\dfrac{\partial V}{\partial T}\right)_p^s$	$\left(\dfrac{p}{T}\right)\left[nC_p\left(\dfrac{\partial V}{\partial p}\right)_T + T\left(\dfrac{\partial V}{\partial T}\right)_p^s\right]$
H	$-V + T\left(\dfrac{\partial V}{\partial T}\right)_p$	nC_p	$nC_p\left(\dfrac{\partial V}{\partial T}\right)_T + T\left(\dfrac{\partial V}{\partial T}\right)_p^s - V\left(\dfrac{\partial V}{\partial T}\right)$	$\dfrac{V_n C_p}{T}$
A	$p\left(\dfrac{\partial V}{\partial p}\right)_T$	$-S - p\left(\dfrac{\partial V}{\partial T}\right)_p$	$-S\left(\dfrac{\partial V}{\partial T}\right)_p$	$\left(\dfrac{1}{T}\right)\left[pnC_p\left(\dfrac{\partial V}{\partial p}\right)_T + pT\left(\dfrac{\partial V}{\partial T}\right)_p^a + TS\left(\dfrac{\partial V}{\partial T}\right)_p\right]$
G	$-V$	$-S$	$-V\left(\dfrac{\partial V}{\partial T}\right)_p - S\left(\dfrac{\partial V}{\partial p}\right)_T$	$\left(-\dfrac{1}{T}\right)\left[nC_p V - TS\left(\dfrac{\partial V}{\partial T}\right)_p\right]$

Table 2.2 Continued

		Constant			
		U	H	A	G
Differential	T	$-T\left(\dfrac{\partial V}{\partial T}\right)_p - p\left(\dfrac{\partial V}{\partial p}\right)_T$	$V - T\left(\dfrac{\partial V}{\partial T}\right)_p$	$-p\left(\dfrac{\partial V}{\partial p}\right)_T$	V
	P	$-nC_p + p\left(\dfrac{\partial V}{\partial T}\right)_p$	$-nC_p$	$S + p\left(\dfrac{\partial V}{\partial T}\right)_p$	S
	V	$-nC_p\left(\dfrac{\partial V}{\partial p}\right)_T - T\left(\dfrac{\partial V}{\partial T}\right)_p^s$	$-nC_p\left(\dfrac{\partial V}{\partial p}\right)_T - T\left(\dfrac{\partial V}{\partial T}\right)_p^s + V\left(\dfrac{\partial V}{\partial T}\right)_p$	$S\left(\dfrac{\partial V}{\partial p}\right)_T$	$V\left(\dfrac{\partial V}{\partial T}\right)_p + S\left(\dfrac{\partial V}{\partial p}\right)_T$
	S	$\left(-\dfrac{p}{T}\right)\left[nC_p\left(\dfrac{\partial V}{\partial p}\right)_T + T\left(\dfrac{\partial V}{\partial T}\right)_p^s\right]$	$\dfrac{VnC_p}{T}$	$\left(-\dfrac{1}{T}\right)\left[pnC_p\left(\dfrac{\partial V}{\partial p}\right)_T + pT\left(\dfrac{\partial V}{\partial T}\right)_p^s + TS\left(\dfrac{\partial V}{\partial T}\right)_p\right]$	$\left(-\dfrac{1}{T}\right)\left[nC_pV - TS\left(\dfrac{\partial V}{\partial T}\right)_p\right]$
	U	0	$V\left[nC_p - p\left(\dfrac{\partial V}{\partial T}\right)_p\right] + p\left[nC_p\left(\dfrac{\partial V}{\partial p}\right)_T + T\left(\dfrac{\partial V}{\partial T}\right)_p^s\right]$	$-p\left[nC_p\left(\dfrac{\partial V}{\partial p}\right)_T + T\left(\dfrac{\partial V}{\partial T}\right)_p^s\right] - S\left[T\left(\dfrac{\partial V}{\partial T}\right)_p + p\left(\dfrac{\partial V}{\partial p}\right)_T\right]$	$V\left[nC_p - p\left(\dfrac{\partial V}{\partial T}\right)_p\right] - S\left[T\left(\dfrac{\partial V}{\partial T}\right)_p + p\left(\dfrac{\partial V}{\partial p}\right)_T\right]$

Differential				
H	$-V\left[nC_p - p\left(\frac{\partial V}{\partial T}\right)_p\right]$ $-p\left[nC_p\left(\frac{\partial V}{\partial T}\right)_p + T\left(\frac{\partial V}{\partial T}\right)_p^s\right]$	0	$\left[S + p\left(\frac{\partial V}{\partial T}\right)_p\right]$ $\times \left[V - T\left(\frac{\partial V}{\partial T}\right)_p - pnC_p\left(\frac{\partial V}{\partial p}\right)\right]$	$VnC_p + VS - TS\left(\frac{\partial V}{\partial T}\right)_p$
A	$p\left[nC_p\left(\frac{\partial V}{\partial p}\right)_T + T\left(\frac{\partial V}{\partial T}\right)_p^s\right]$ $+ s\left[T\left(\frac{\partial V}{\partial T}\right)_p + p\left(\frac{\partial V}{\partial p}\right)_T\right]$	$-\left[s + p\left(\frac{\partial V}{\partial T}\right)_p\right]$ $\times \left[V - T\left(\frac{\partial V}{\partial T}\right)_p - pnC_p\left(\frac{\partial V}{\partial p}\right)_T\right]$	0	$-S\left[V + p\left(\frac{\partial V}{\partial p}\right)_T\right] + pV\left(\frac{\partial V}{\partial T}\right)_p$
G	$-V\left[nC_p - p\left(\frac{\partial V}{\partial T}\right)_p\right]$ $+ s\left[T\left(\frac{\partial V}{\partial T}\right)_p + p\left(\frac{\partial V}{\partial p}\right)_T\right]$	$-VnC_p - VS + TS\left(\frac{\partial V}{\partial T}\right)_p$	$S\left[V + p\left(\frac{\partial V}{\partial p}\right)_T\right] + pV\left(\frac{\partial V}{\partial T}\right)_p$	0

Table 2.3 Viscosity Conversion.

	Klasmatic	
To convert from	**To**	**Multiply by**
cm²/sec (Stokes)	Centistokes	10^2
	ft²/hr	3.875
	ft²/sec	1.076×10^{-2}
	in.²/sec	1.550×10^{-1}
	m²/hr	3.600×10^{-1}
cm²/sec × 10² (Centistokes)	cm²/sec (Stokes)	1×10^{-2}
	ft²/hr	3.875×10^{-2}
	ft²/sec	1.076×10^{-2}
	in.²/sec	1.550×10^{-2}
	m²/hr	3.600×10^{-2}
ft²/hr	cm²/sec (Stokes)	2.581×10^{-1}
	cm/sec × 10² (Centistokes)	2.581×10
	ft²/sec	2.778×10^{-4}
	in.²/sec	4.00×10^{-2}
	m²/hr	9.290×10^{-3}
ft²/sec	cm²/sec (Stokes)	9.29×10^3
	cm²/sec × 10³ (Centistokes)	9.29×10^4
	ft²/hr	3.60×10^3
	in.²/sec	1.44×10^2
	m²/hr	3.345×10^3
in.²/sec	cm²/sec (Stokes)	6.452
	cm²/sec × 10² (Centistokes)	6.452×10^2
	ft²/hr	2.50×10
	ft²/sec	6.944×10^{-2}
	m²/hr	2.323
m²/hr	cm²/sec (Stokes)	2.778
	cm²/sec × 10³ (Centistokes)	2.778×10^2
	ft²/hr	1.076×10
	ft²/sec	2.990×10^{-3}
	in.²/sec	4.306×10^{-1}

Table 2.3 *Continued*

Absolute viscosity = kinematic viscosity × density; lb = mass pounds; lb_F = force pounds
Absolute

To convert from	To	Multiply by
gm/(cm)(sec) [Poise]	gm/(cm)(sec)(10^2) [Centipoise]	10^2
	kg/(m)(hr)	3.6×10^2
	lb/(ft)(sec)	6.72×10^{-2}
	lb/(ft)(hr)	2.419×10^2
	lb/(in.)(sec)	5.6×10^{-3}
	(gm_F)(sec)/cm^2	1.02×10^{-2}
	(lb_F)(sec)/in.^{2} [Reyn]	1.45×10^{-6}
	(lb_F)(sec)/ft^2	2.089×10^{-2}
gm/(cm)(sec)(10^2) [Centipoise]	gm/(cm)(sec) [Poise]	10^{-2}
	kg/(m)(hr)	3.6
	lb/(ft)(sec)	6.72×10^{-4}
	lb/(ft)(hr)	2.419
	lb/(in.)(sec)	5.60×10^{-5}
	(gm_F)(sec)/cm^2	1.02×10^{-5}
	(lb_F)(sec)/in.^{2} [Reyn]	1.45×10^{-7}
	(lb_F)(sec)/ft^2	2.089×10^{-4}
kg/(m)(hr)	gm/(cm)(sec)	2.778×10^{-3}
	gm/(cm)(sec)(10^2) [Centipoise]	2.778×10^{-1}
	lb/(ft)(sec)	1.867×10^{-4}
	lb/(ft)(hr)	6.720×10^{-1}
	lb/(in.)(sec)	1.555×10^{-4}
	(gm_F)(sec)/cm^2	2.833×10^{-4}

number (R_{rot}), the Prandtl number (P_r), the pseudo-Reynolds number (R_{ep}), and the eccentricity parameter (ε) [2].

2.3.1 LINEAR HEAT FLOW

Rotating system geometry and its coordinates are shown in Figure 2.1. For a rotating system, the ratio of coefficient of frictional resistance under rotation to coefficient of frictional resistance under static condition is

Table 2.4 Important Properties of Cryogenic Gases.

Gas	BP@1atm pressure (K)	MP@ 1atm pressure (K)	Heat of vapor @ BP (J/g)	Critical temperature (K)
Helium HE^3	3.2	1.0	—	—
Helium HE^4	4.2	2.0	20.5	4.183
Nitrogen	77.37	63.4	199.0	126.1
Freon-14	145.14	89.5	134.8	227.5
Freon-22	232.5	113.0	235.0	369.0
Freon-12	243.1	118.0	167.2	384.0
Freon-11	296.8	162.7	237.3	471.0

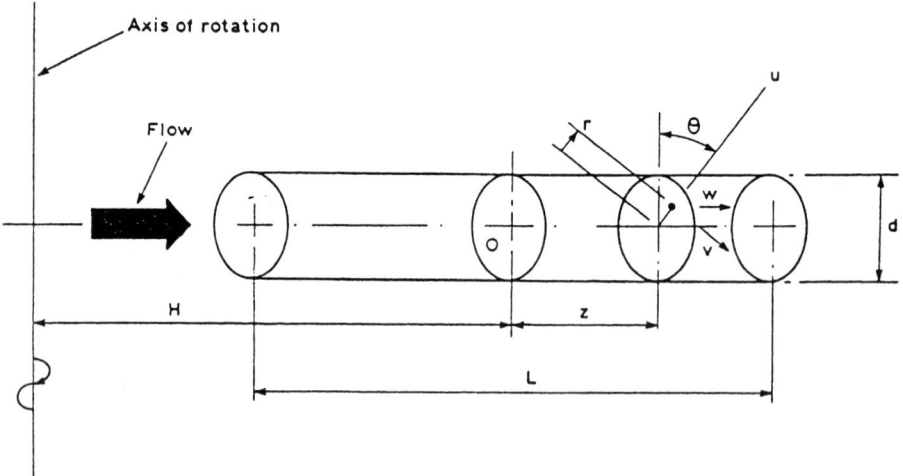

Fig. 2.1 Cordinates of a rotating system with respect to the axis of rotation and flow direction.

known as the frictional coefficient ratio (FCR). This FCR parameter is dependent on the rotational Reynolds number (J_a) and through flow pseudo-Reynolds number (R_{ep}) and can be expressed as

$$\text{FCR}(J_a, R_{ep}) = \left[1 - J_a^2 \left(1/576 + R_{ep}^2/1{,}032{,}192\right)\right]^{-1} \quad (2.1)$$

where J_a is the rotational Reynolds number, which is strictly dependent on axial pressure.

2. Heat Flows on Heat Exchanger Performance and Cooler Efficiency   35

Calculated values of FCR as a function of the J_a and R_{ep} variables are shown in Figure 2.2(A) through Figure 2.2(F). The parameter J_a varies from 1 to 16, while the parameter R_{ep} varies from 10 to 60. It is evident from these figures that the magnitude of FCR regains the negative values when the parameter J_a exceeds a value of 16 and when R_{ep} exceeds a value

Sample Calculation showing Computation of the frictional coefficient ratio (FCR) defined by $\dfrac{C_{fR}}{C_{fo}}$ Ratio

$J_a := 1, 2 .. 16$

$R_{ep} := 10, 20 .. 50$

$$FCR(J_a, R_{ep}) := \left[1 - J_a^2 \cdot \left(\frac{1}{576} + \frac{R_{ep}^2}{1032192} \right) \right]^{-1}$$

$FCR(J_a, 10) =$	$FCR(J_a, 20) =$	$FCR(J_a, 30) =$	$FCR(J_a, 40) =$	$FCR(J_a, 50) =$
1.002	1.002	1.003	1.003	1.004
1.007	1.009	1.011	1.013	1.017
1.017	1.019	1.024	1.03	1.039
1.03	1.035	1.044	1.055	1.071
1.048	1.056	1.07	1.09	1.116
1.071	1.083	1.104	1.134	1.176
1.099	1.116	1.147	1.192	1.256
1.133	1.157	1.2	1.266	1.363
1.174	1.208	1.268	1.363	1.508
1.224	1.27	1.353	1.489	1.712
1.285	1.346	1.461	1.66	2.013
1.359	1.441	1.601	1.898	2.492
1.449	1.56	1.788	2.249	3.364
1.561	1.713	2.046	2.81	5.405
1.702	1.915	2.42	3.837	15.524
1.884	2.191	3.009	6.3	-15.508

Fig. 2.2

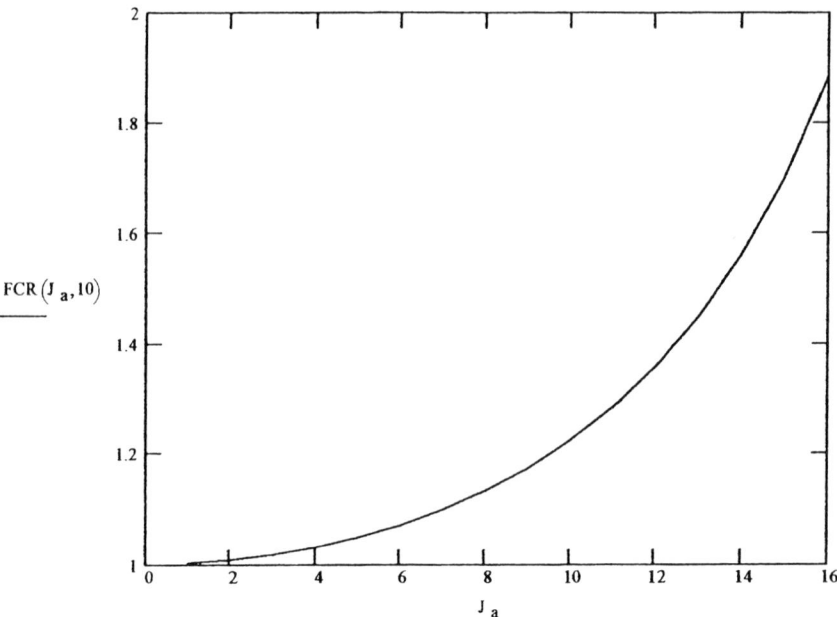

Fig. 2.2(A) Frictional coefficient ratio (FCR) for $R_{ep} = 10$.

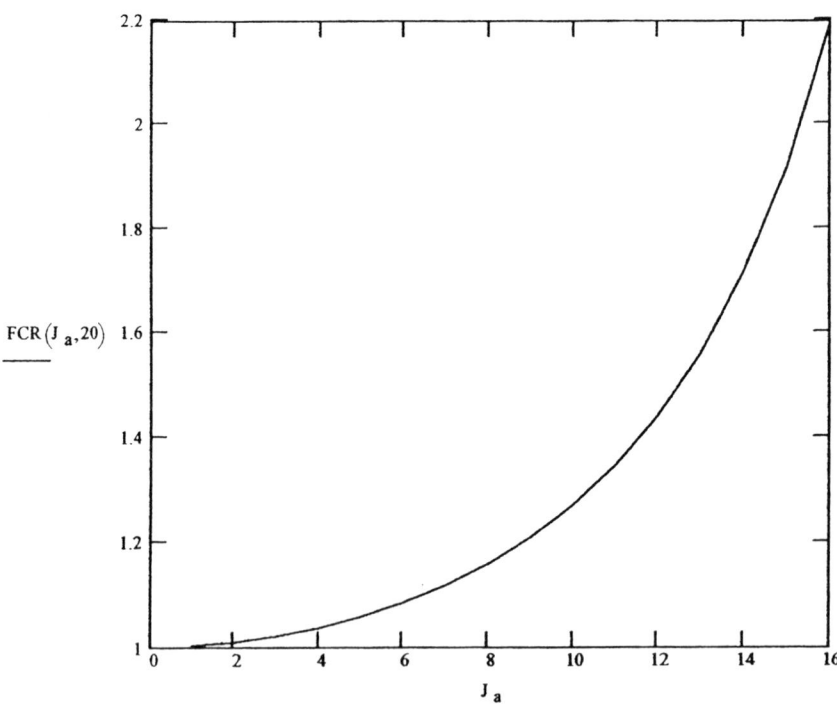

Fig. 2.2(B) Frictional coefficient ratio (FCR) for $R_{ep} = 26$.

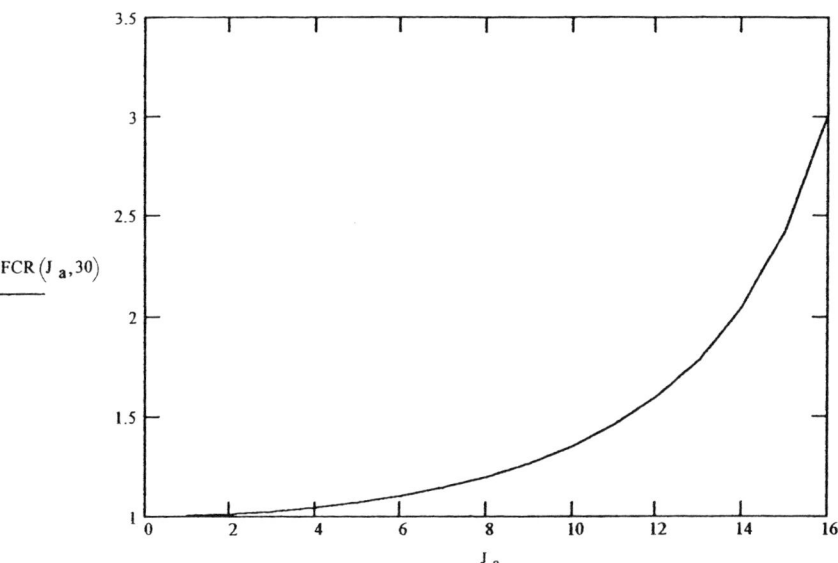

Fig. 2.2(C) Frictional coefficient ratio (FCR) for $R_{ep} = 30$.

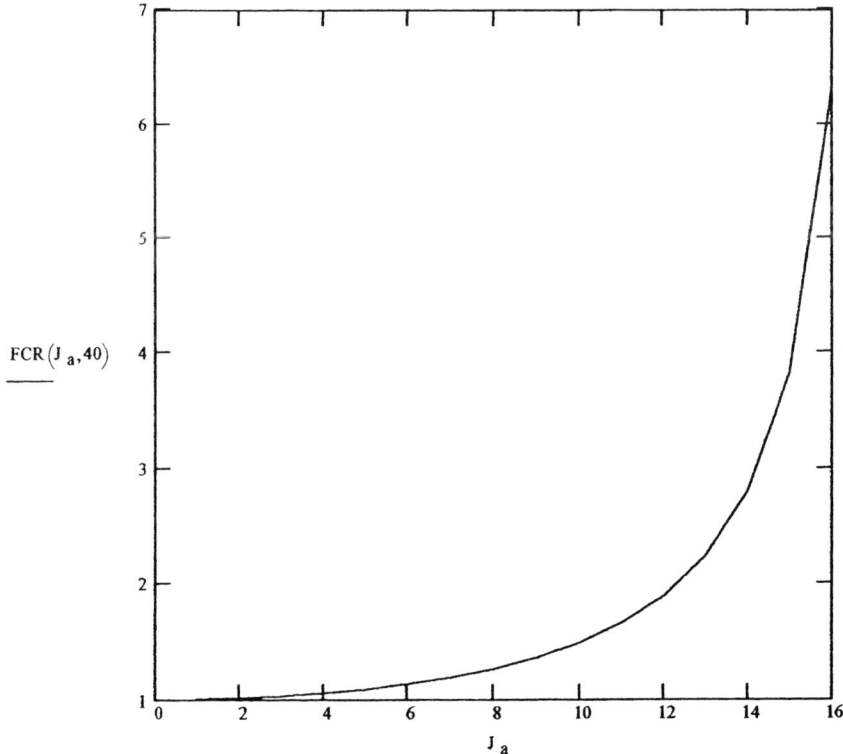

Fig. 2.2(D) Frictional coefficient ratio (FCR) for $R_{ep} = 40$.

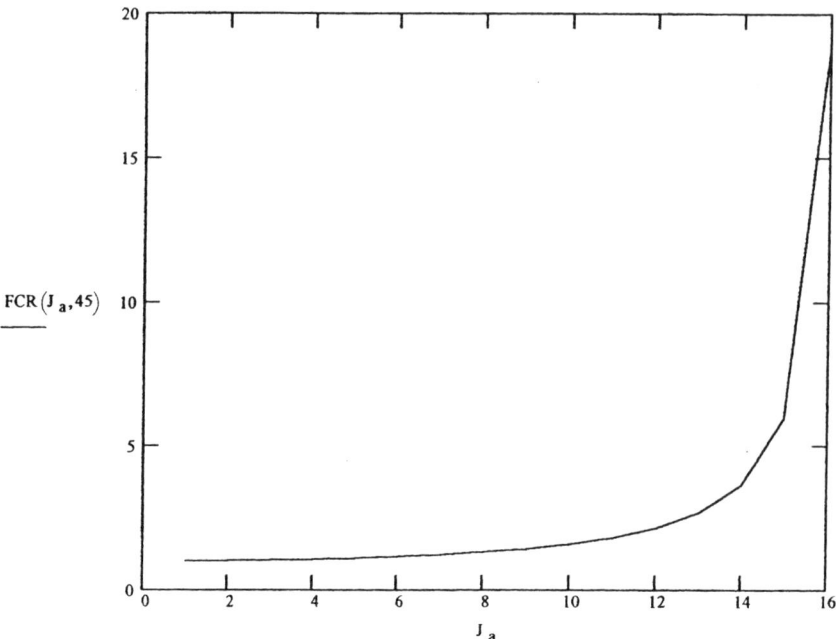

Fig. 2.2(E) Frictional coefficient ratio (FCR) for $R_{ep} = 45$.

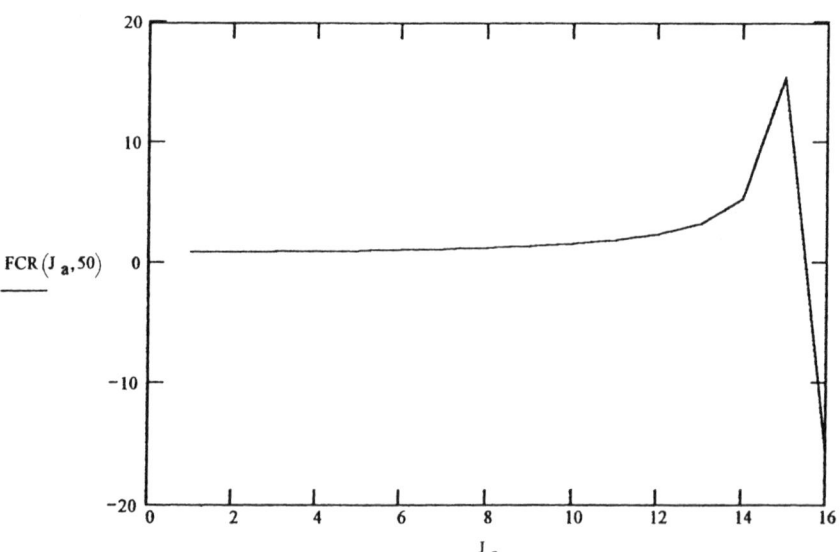

Fig. 2.2(F) Frictional coefficient ratio (FCR) for $R_{ep} = 50$.

2. Heat Flows on Heat Exchanger Performance and Cooler Efficiency

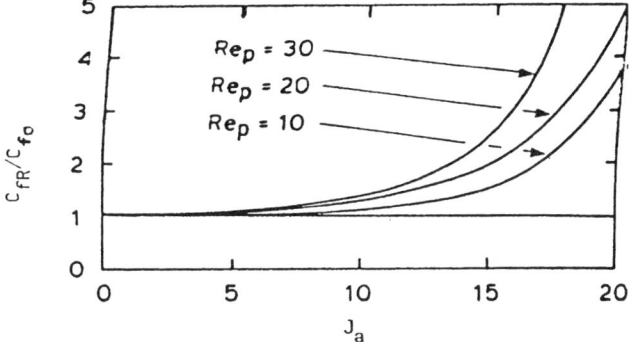

Fig. 2.3 Impact on rotation on the flow resistance ratio (C_{FR}/C_{fo}) where C_{FR} is the coefficient flow with resistance and C_{fo} is the coefficient without resistance.

of 50. The impact of rotation on the overall flow resistance for three distinct values of parameter is illustrated in Figure 2.3.

In case of low rotational speed with respect to the mean flow along the axis, the entire flow regime is strictly controlled by inertia and pressure in the boundary layer region near the tube wall. Studies performed by the author on fluid flows through the tubes indicate the secondary flow components in the core regions are likely to be significantly smaller than those in the axial direction. When pressure-related terms are eliminated from the flow equation, it can be shown that the core-region axial field vorticity field is functionally related to $R\sin\theta$, which is the projection of the position vector of a typical point in the fluid flow as viewed along the z-axis and as illustrated in Figure 2.1. The axial velocity of the flow in the core region will have its components related to a stream function.

It is important to mention that a laminar flow in a circular or square pipe or duct has an impact on the parameter FCR, which is defined as the ratio of coefficient of friction due to rotation to coefficient of friction with no rotation. This can be expressed as C_{fR}/C_{fo}, where R stands for rotation and o stands for no rotation or static position. Equation (2.1) determines the impact of rotation on the mean flow through a cooling pipe, which is dependent on the mean velocity component in the axial direction. Note that the mean velocity is usually related to the through flow Reynolds number (Re). In the case of laminar flow, as the heat flux increases, the local heat transfer rate also increases. This reflects the complex interaction between the Coriolis acceleration effect in the entry region with the centripetal buoyancy under the rotating environments [2].

2.3.1.1 Impact of Linear Flow on Heat Exchanger Performance

The heat transfer in a heat exchanger composed of tubes with specific cross sections is dependent on flow resistance with unheated flow, flow resistance with heated fluid flow, the heat transfer component associated with laminar flow, and the heat transfer component contributed by turbulent flow, if any. Studies performed by the author on fluid flow characteristics indicate that fluid flow and heat transfer in circular tubes that rotate about an orthogonal axis (as shown in Figure 2.1) may not be suitable for certain rotating system applications.

For low rotating speed systems, the axial pressure-type effect cannot be ignored. However, for high rotating speeds at which the rotational Reynolds number (J_d) is much greater than the through-flow Reynolds number (Re), the inertial effects are negative in the core and boundary layer regions. For a high-speed rotating system, one can express the FCR ratio as

$$\text{FCR} = [C_{fR}/C_{fo}] = \left[(0.0902)/(J_d Re/\Gamma)^{1/4}\right]/\left[1 - 3.354/(J_d Re/\Gamma)^{1/4}\right] \quad (2.2)$$

where

$$\Gamma = \left[1 + (0.3125)(J_d/Re)^2\right]^{1/2} - [(0.56)(J_d/Re)] \quad (2.3)$$

Γ represents the pressure load parameter that affects the magnitude of FRC.

In cases where $J_d \gg Re$, Equation (2.2) is reduced to

$$\text{FCR} = [(0.0675\sqrt{J_d})]/[1 - (2.11)\sqrt{J_d}] \quad (2.4)$$

Note that the function FCR has been calculated assuming that the through Reynolds number Re is much greater than the rotational Reynolds number J_d or J_d is much greater than Re. Calculated values of this function are shown in Figure 2.4 when J_d is much greater than Re. This function is represented by the first two curves illustrated in Figure 2.5, when Re is much greater than J_d. The impact of rotation on the laminar flow resistance and comparison between flow resistance magnitudes under various

2. Heat Flows on Heat Exchanger Performance and Cooler Efficiency

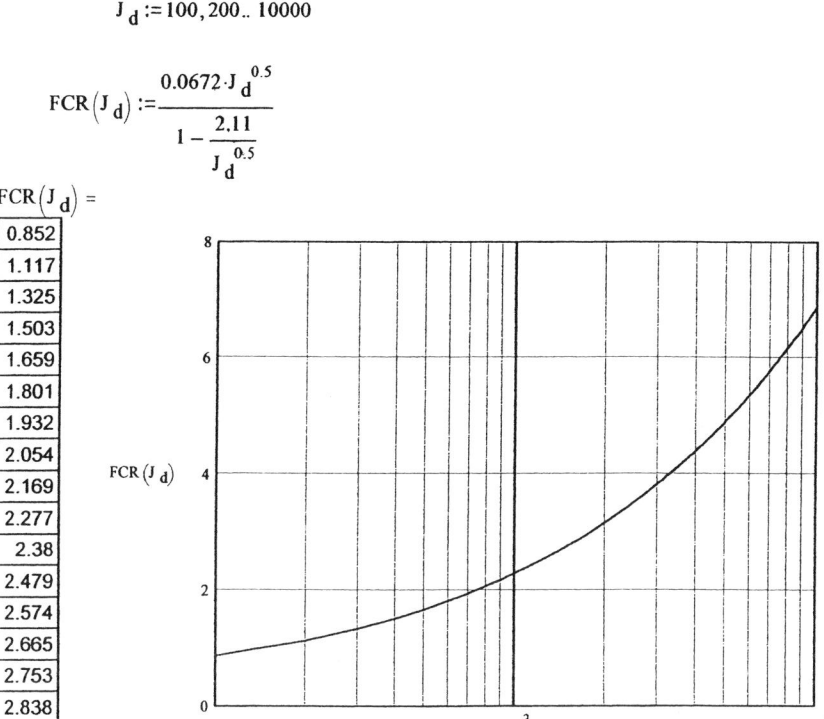

Fig. 2.4 Impact of rotation on the laminar flow resistance when the parametr $J_d \gg R_e$.

assumed parametric conditions is shown in Figure 2.5. The influence of rotation on mean flow rate [2], as a function of the through-flow Reynolds number Re and pseudo-Reynolds number Rep, is evident from Figure 2.6.

2.3.2 IMPACT OF TURBULENT FLOW ON HEAT EXCHANGER PERFORMANCE

In case of turbulent heat flow, two types of heat flow are of critical importance: flow with parallel-mode rotation and flow with orthogonal-mode

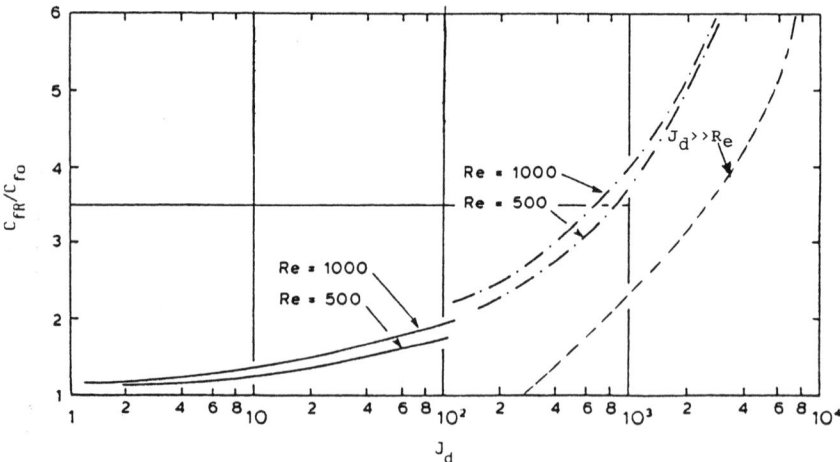

Fig. 2.5 Impact of rotation of the C_{FR}/C_{fo} ratio when parameter $J_d < R_e$, when $J_d > R_e$, and when $J_d \gg R_e$.

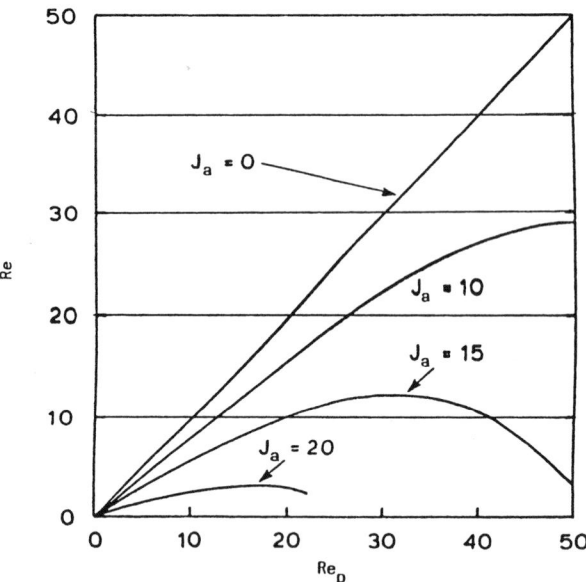

Fig. 2.6 Impact of rotation on the mean flow rate as a function of through-flow Reynolds number (R_e) and pseudo-Reynolds number (R_{ep}) for various values of parameter J_a.

rotation. Under turbulent heat flow in a circular tube or duct, significant enhancement in heat transfer due to Coriolis acceleration has been observed, particularly with the orthogonal mode of rotation. When the fluid has a Prandtl (P_r) number greater than unity, the ratio of Nusselt (N_u) under rotation and static conditions [2] can be written as

$$[N_{u\infty}/N_{uo}] = [(1.07)(X)^{0.33}][1 + 0.059/(X_o)^{.166}] \tag{2.5}$$

where

$$X = [J_d/4Re\Gamma^2] \tag{2.6}$$

and

$$\Gamma = [1 + 1.285(J_d/Re)^2] - [1.135(J_d/Re)] \tag{2.7}$$

The zero speed Nusselt number (N_{uo}), when the Pr number is greater than unity, can be written as

$$N_{uo} = [0.034(Re)^{0.75}] \tag{2.8}$$

When the Prandtl number (P_r) is less than unity, Equations (2.6), (2.7), and (2.8) are simplified and can be rewritten as

$$X = [0.177/(Re)^{1.5}][J_d/\Gamma]^{2.5} \tag{2.6a}$$

$$\Gamma = [1.162(J_d/Re)^2]^{0.5} - [1.25(J_d/Re)] \tag{2.7a}$$

$$N_{uo} = [(0.023)(Re)^{0.8}(P_r)^{0.4}] \tag{2.8a}$$

These equations illustrate the impact of rotation on turbulent heat transfer as a function of the through Reynolds number (Re), Prandtl number (P_r), and rotational Reynolds number (J_d). The ratio $N_{u\infty}/N_{uo}$ can be treated as a normalized Nusselt number and is simply represented by a new symbol Nu, which is a function two variables: Re and J_d. The author has used the Math Cad software program to compute the values of the parameter Nu as a function of several variable parameters involved when the Prandtl number (P_r) is greater than unity and less than unity. Computed values of the new parameter Nu and the impact of rotation on turbulent heat transfer are illustrated in Figure 2.7, when the Prandtl

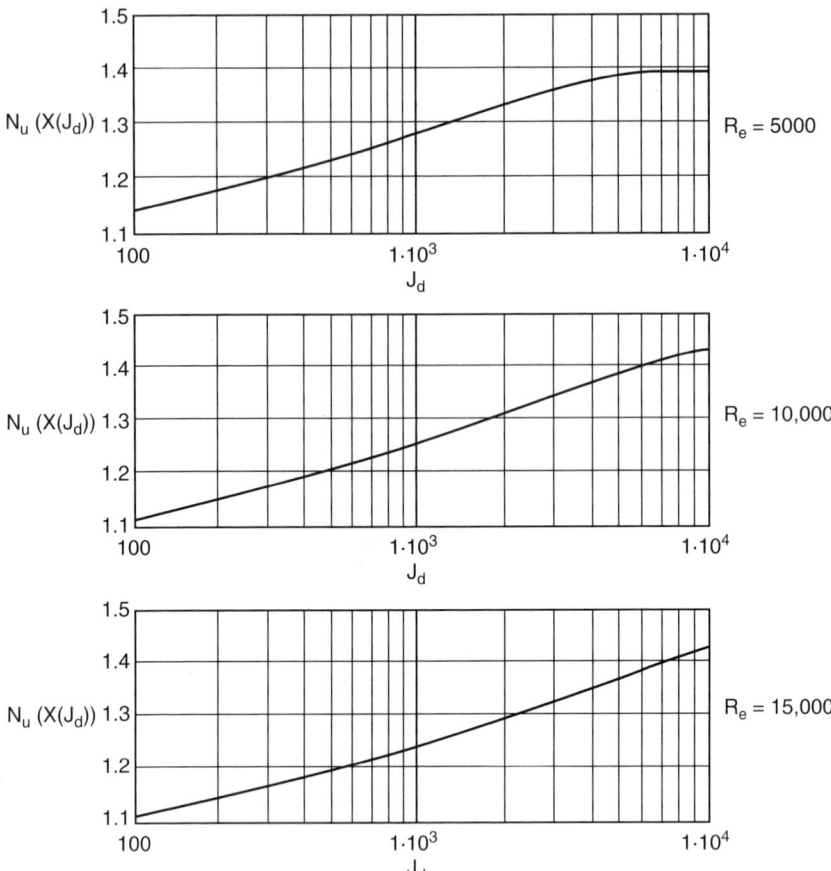

Fig. 2.7 Impact of rotation on the turbulent flow as a function of parameter R_e and J_d when the Prandtl number (P_r) is greater than 1.0.

number is greater than unity. The impact of rotation on turbulent heat transfer when the Prandtl number is less than unity is shown in Figure 2.8. The effects of variables X and J_d on the overall FCR and Nu parameters and the impact of rotation on turbulent heat transfer for two extreme values of the Prandtl number are illustrated in Figure 2.9.

Based on these computations, it can be concluded that the Coriolis effect produces significant improvement in heat transfer; nevertheless the enhancement is not as high as with laminar flow. The rotation plots shown in various figures indicate that for specific values of the rotational

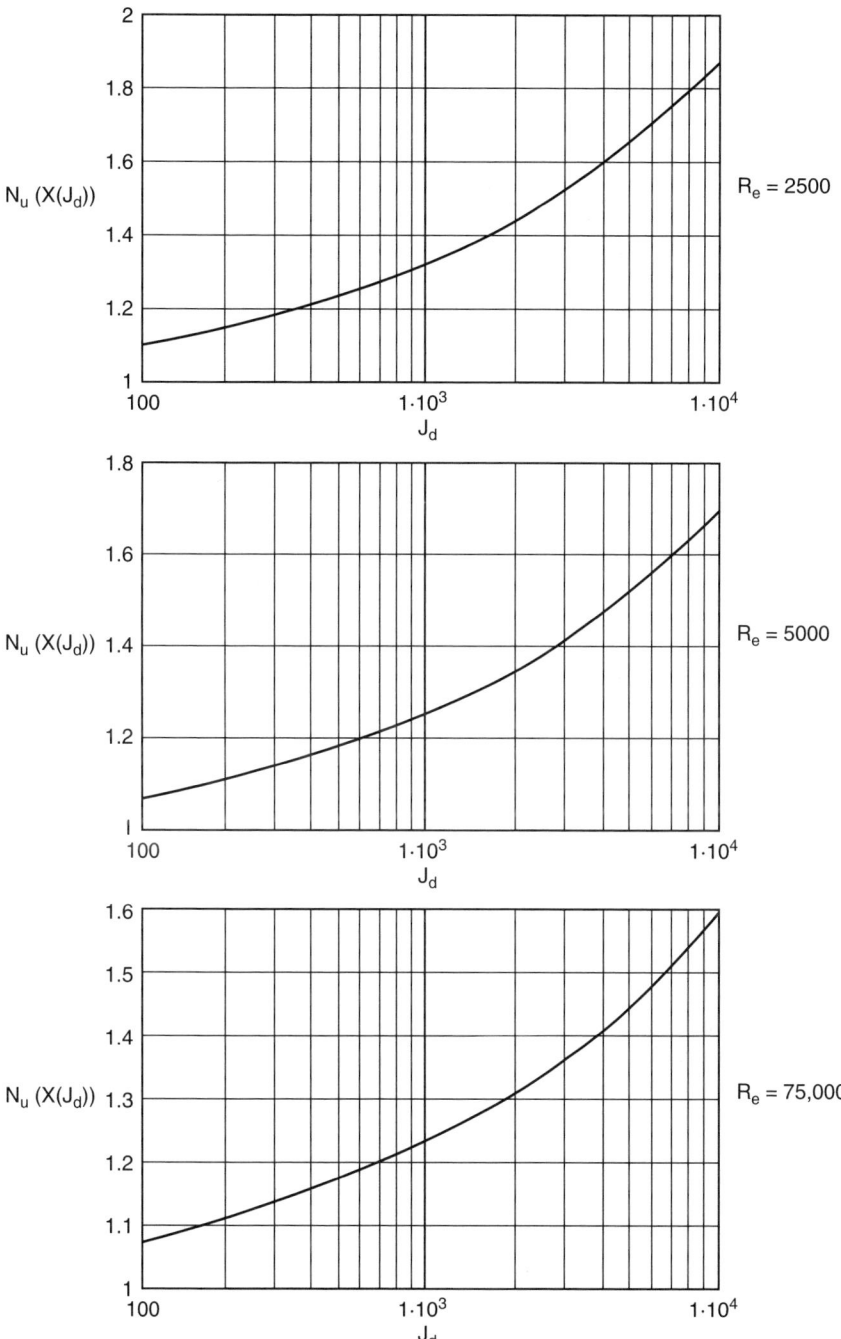

Fig. 2.8 Impact of rotation on the turbulent flow normalized Nusselt number (N_u) as a function of parameters R_e and J_d when $P_r < 1$.

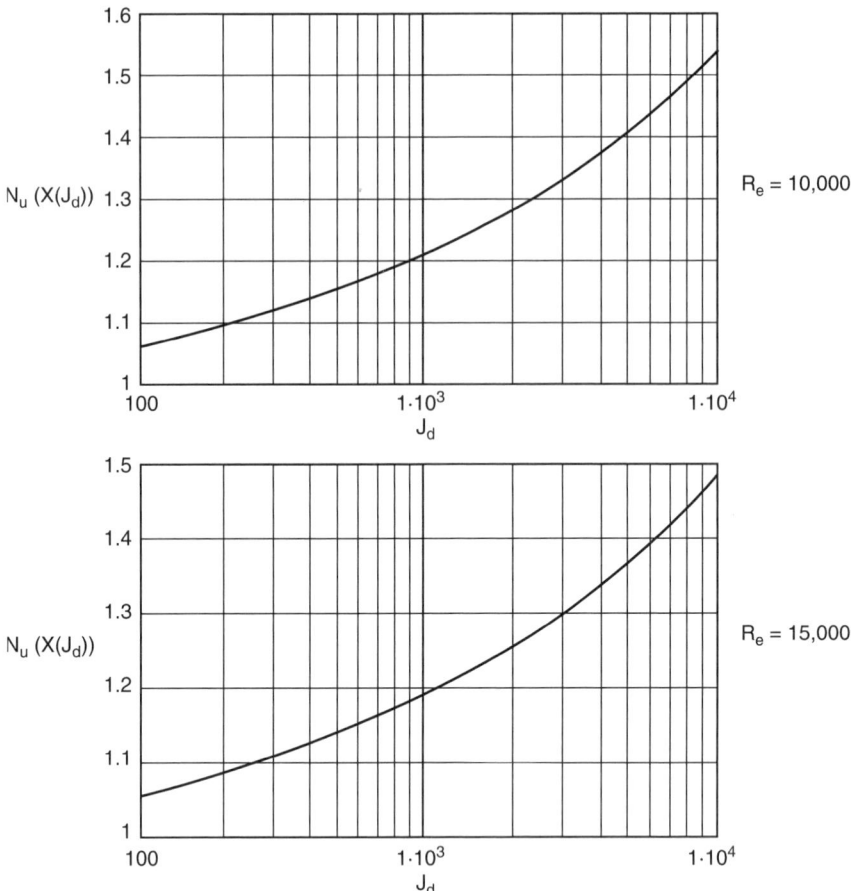

Fig. 2.8 (*Continued*)

Reynolds number, the enhancement decreases as the through-flow Reynolds number increases. The plots shown in Figure 2.9 reveal the dependence of turbulent flow on heat and mass transfer parameters. It is important to mention that the circumferential variation of Nusselt number (*Nu*) is measured from the mass transfer experimental tests to preserve high accuracy. Furthermore, the calculations on turbulent flow indicate that the angular variation is not so severe as that found with laminar flow. The impact of rotation under turbulent flow is strictly dependent on heated radially outward flow.

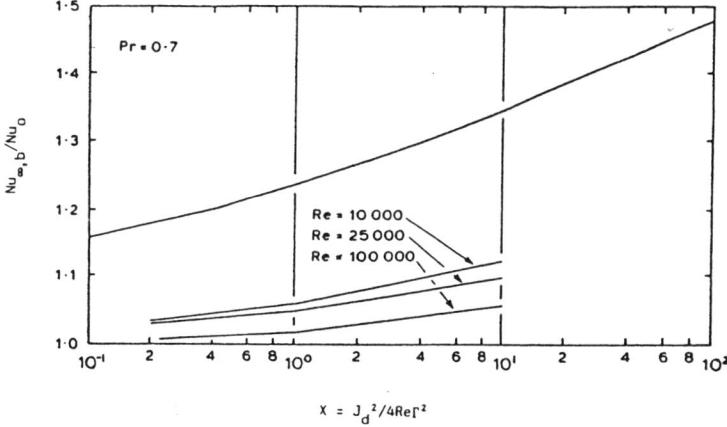

Normalized Nusselt number as a function of parameter X, J_d, and R_e when P_r is less than unity.

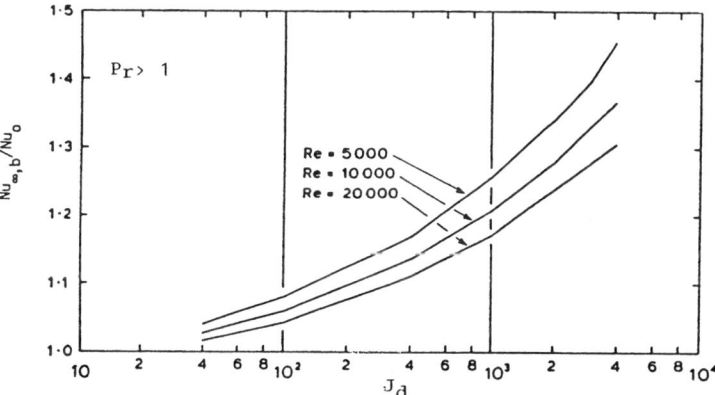

Normalized Nusselt number as a function of R_e and J_d when P_r is > 1

Fig. 2.9 Normalized Nusselt number for a turbulent heat flow as a function of variables R_e, J_d, and P_r.

2.4 Two-Dimensional Heat Flow Model

A numerical model is required to study the unique features of gas flow and heat transfer in the cryogenic fluid known as superfluid helium. The simplest model based on the original two-fluid model involves a conventional continuity equation, a momentum equation for the total fluid, and an energy equation in the form of the temperature-based energy equation in which the heat flux due to Gorter-Mellink [3] internal convection is properly incorporated. One-dimensional and two-dimensional heat transfer equations must be considered to realize the most accurate results for the model. To demonstrate the full capability of the original model in solving the multidimensional problems, two-dimensional analysis must be performed for the internal-convection heat transfer case with one of the walls partially heated.

To understand the impact of heat flow and heat transfer efficiency in a heat exchanger, one requires a clear understanding of the first law of thermodynamics, which states that heat input (Hi) is equal to heat out (Ho). The law is expressed in terms of various parameters and can be written as

$$Hi = [(dm/dt)Cp(T_2 - T_1)] + [Ho] \qquad (2.9)$$

where

$$[dm/dt] = [(Hi - Ho)/Cp\Delta T] \qquad (2.10)$$

and dm/dt is the mass flow rate, Cp is the specific heat of the fluid, and T_2 and T_1 are the temperature at the inlet and outlet of the tube carrying the fluid.

These temperatures will be different under laminar and turbulent flow environments because of variations in internal heat generated. For a laminar flow, the mass flow through a tube is inversely proportional to the temperature difference (ΔT) as is evident from Equation (2.10).

As stated earlier, a transient two-dimensional analysis [3] must be performed based on the internal-convection heat transfer in a helium II (He II) pool with one of the walls partially heated to achieve optimum results. A modified model involves approximating the momentum equations for the superfluid and the normal-fluid components in the original two-fluid model by eliminating the two large contributions from the thermomechanical and the Gorter-Mellink mutual friction effects. Note that the contributions from these two effects are larger than other terms by a few orders of magnitude, which can make the numerical analysis more complex and can introduce numerical instability. The original two-fluid model plays an important role in the derivation of the simplified model, which is briefly described next.

2.4.1 DESCRIPTION OF MODIFIED TWO-FLUID MODEL

The two assumed imaginary components of helium II include the nonviscous superfluid component in which the rotation of the velocity vector is equal to zero and the viscous-normal fluid component that is responsible for total energy transport. The density (d) of total fluid can be defined as

$$d = [d_n + d_s] \tag{2.11}$$

where the subscripts n and s represent normal-fluid and superfluid components, respectively.

The density flux equation can be written as

$$d\mathbf{v} = [d_n \mathbf{v}_n + d_s \mathbf{v}_s] \tag{2.12}$$

where $\mathbf{v}$ is the velocity vector of the total fluid and the subscripts n and s denote the normal-fluid velocity vector and superfluid velocity vector components, respectively.

Neglecting the larger terms, a good approximation for the expression of a modified two-fluid model can be written as

$$[E\nabla T] = \left[-Pd_n |\mathbf{v}_n - \mathbf{v}_s|^2 (\mathbf{v}_n - \mathbf{v}_s)\right] \tag{2.13}$$

where P is the Gorter-Mellink parameter, E is the entropy, and ∇ is the Del operator.

The two velocity vector components can be written as

$$\mathbf{v}_n = \left[\mathbf{v} - (d_s^3 E/A)^{0.33} (\nabla T)\right] \tag{2.14}$$

$$\mathbf{v}_s = \left[\mathbf{v} + (d_n^3 E/A)^{0.33} (\nabla T)\right] \tag{2.15}$$

where T is the absolute temperature, and A is a complex variable, which is defined next:

$$A = \left[Pd^3 d_n |\nabla T|^2\right] \tag{2.16}$$

where P is the Gorter-Mellink parameter, and d is the density variable. It is important to mention that the square of the product (∇T) expresses a dyadic product of two identical vectors ∇T. This is a second-order tenser whose (i, j) component is expressed as [($\partial T/\partial x_i$)($\partial T/\partial x_j$)]. The convectional acceleration and the viscous effects are caused by thermomechanical force.

The energy (U) equation can be written as

$$U = [dC_p \partial T/\partial t] = [-dC_p(\mathbf{v} \cdot \nabla)T] - [\nabla \cdot \{d_s ET(\mathbf{v}_n - \mathbf{v}_s)\}] \quad (2.17)$$

where C_p is the specific heat of the medium at constant pressure. The second term on the right side of Equation (2.17) represents the heat flux of the internal convection (Qi), which can be defined as

$$Qi = [d_s E(\mathbf{v}_n - \mathbf{v}_s)] \quad (2.18)$$

Inserting Equation (2.18) into Equation (2.13), one gets

$$Qi = \left[-(1/f(T))^{0.33} \right] \left[\nabla T / |\nabla T|^{0.67} \right] \quad (2.19)$$

where $f(T) = [Pd_n/(d_s^3 E^4 T^3)]$ and is called the heat conductivity function. It is equally important to mention that the one-dimensional expression of Equation (2.19) is known as the Gorter-Mellink power law equation [3]. Using the previous equations, one can now obtain one-dimensional transient temperature profiles as a function of time, temperature, distance for a given value of heat flux of internal convection, and initial temperature. An analytical one-dimensional system shown in Figure 2.10 consists of a vertical Gorter-Mellink tube or duct with a height of 15 cm with the heat flux at the bottom and the temperature at the top end [3]. Transient temperature profiles using the simplified model and the two-fluid model for the one-dimensional system are shown in Figure 2.10. The temperature at the heated section of the system shown in Figure 2.10 can be computed by means of the temperature gradients given by Equation (2.19). The same equation can be modified for a two-dimensional system involving horizontal and vertical directions represented by x and z ordinates, respectively. Thus, the heat flux can be obtained by modifying Equation (2.18) for a two-dimensional system and can be written as

$$Qi(z) = \left[-(1/f(T))^{0.33} \right] [\partial T/\partial z] \left[(\partial T/\partial x)^2 + (\partial T/\partial z)^2 \right]^{-0.33} \quad (2.20)$$

where x and z indicate the horizontal and vertical directions, respectively. This expression assumes that the adiabatic condition is applied elsewhere on the wall, the temperature gradient is constant at the top of the wall

2. Heat Flows on Heat Exchanger Performance and Cooler Efficiency

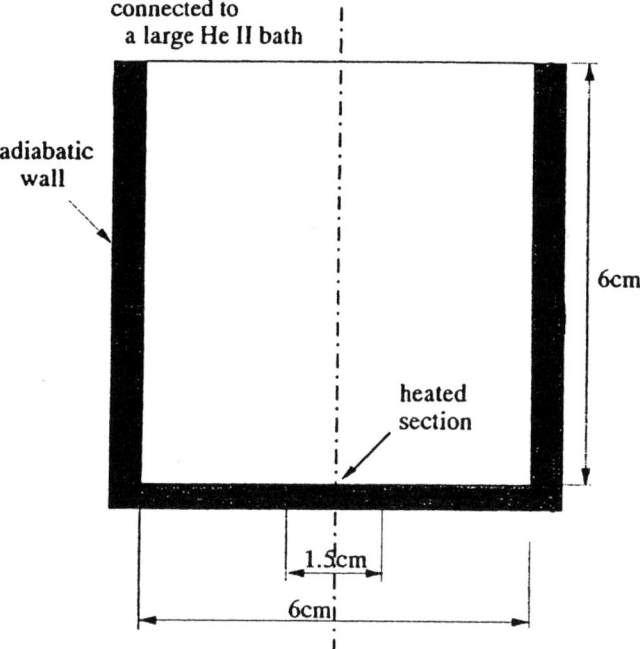

Fig. 2.10 One-dimensional system showing the heated section and the adiabatic wall portion for analytical modeling.

shown in Figure 2.10, and the pressure at the top of the boundary is constant. The temperature distribution is dependent on the boundary conditions and the relaxation time for the temperature, which can vary between 0.1 second to 2 seconds. The fast relaxation of temperature is strictly due to the large thermal conductivity of the He II. Studies performed by the author on the superfluid and normal-fluid velocity components indicate that a vortex of the total fluid can develop near the junction between the heated and adiabatic boundaries after a short relaxation period not exceeding 0.1 second. The studies further indicate the vortex is not generated by the buoyancy-induced natural convection but by the thermomechanical force of the He II refrigerant. The three maximum velocities — the superfluid, normal-fluid, and total fluid velocities — are generally used to monitor the development of the flow pattern toward a steady-state heat flux profile.

In summary, a simplified version of original two-fluid model is capable of performing multidimensional numerical analyses on heat transfer and fluid flow in He II medium or any refrigerant. It is interesting to note that the original two-fluid model has been proved successful for both the one-dimensional problem and two-dimensional analysis. The two-fluid model is best suited for thermal analysis of space-based cryogenic cooling systems.

2.5 Heat Transfer Rates for Heat Exchangers

The theoretical aspects of heat transfer are of fundamental importance in the design and development of cryogenic coolers for specific applications. The transfer of heat can occur in three distinct ways: conduction, radiation, and convection. Heat transfer rate (dQ/dt) in a solid, a liquid, or air depends on the surface area, thermal conductivity, boundary temperatures, emissivity factor, configuration factor, and other constants involved. A brief description of the three modes of heat transfer is provided, with emphasis on critical operating parameters.

2.5.1 CONDUCTION MODE OF HEAT TRANSFER

The fundamental equation for the heat transfer rate or heat flux across two surfaces through conduction can be written as

$$[dQ/dt] = [KA(T_2 - T_1)/L] \text{W/sec} \qquad (2.21)$$

where K is thermal conductivity (W/cm/K), A is the surface area (cm^2) normal to the direction of heat flow, T_2 and T_1 are the two surface temperatures (K), and L is the thickness of the material (cm). This is a case of heat transfer from one metal to another or from one surface to another. In the case of heat conduction across the interfaces, the ratio (K/L) can be replaced by a joint interface conductance, which is a function of thermal conductivity, surface finish, surface hardness, surface flatness, mean temperature, contact pressure, and the presence of oxide film or interface shim materials such as brass foil or lead foil or gas. A review of published literature indicates that the thermal interface conductance for various metal pairs composed of stainless steel, copper, iron, or tungsten varies between 0.2 to 1.0 W/cm/K over the contact pressure range from 10 to 200 p.s.i. However, its value for aluminum with vacuum or air as a gap material is extremely low and can vary from

0.02 to 0.08 W/cm/K over the contact pressure ranging from 5 to 40 p.s.i. [4].

2.5.2 CONVECTION MODE OF HEAT TRANSFER

Convection process involves the heat transfer from a fluid flowing over a hot surface or cold surface. When the circulation is produced due to density variations as a result of temperature changes, the convection is called "free" convection. When the circulation is made by some mechanical means involving a pump, fan, or relative motion of the fluid, the convection is known as "forced" convection. When a fluid flows over a hot or cold surface, a stagnant thin film adheres to the surface, which acts like a barrier to heat transfer. Under these circumstances, heat is conducted through this film and is dependent on the size and shape of the surface, film thickness, fluid viscosity, the specific heat and thermal conductivity of the fluid, the velocity of the fluid stream, and the temperature difference on the two sides of the film. The rate of heat flow through this film can be expressed as

$$[dQ/dt] = [H_c A(T_s - T_f)] \text{W/sec} \qquad (2.22)$$

where H_c is the heat transfer coefficient of the film (W/cm^2/K), A is surface area (cm^2), T is the absolute temperature (K), and the subscripts s and f stand for surface and film, respectively. For unbounded convection involving plates, tubes, or bodies of revolution immersed in fluid, it is desirable to define the film temperature of the fluid far away from the surface, which may be significantly higher. For bounded convection involving fluids flowing in tubes or channels, the film temperature must be taken as the enthalpy-mixed mean temperature. However, in a convection process involving high-velocity gases as in the case of aircraft engines, it is necessary to replace the film temperature with a recovery temperature, which is also known as the adiabatic wall temperature. The recovery temperature or the adiabatic wall temperature [4] can be written as

$$T_{aw} = T_v \left[1 + (\gamma - 1)/2(M_f)^2\right] \qquad (2.23)$$

where γ is the ratio of the specific heats at constant pressure and constant volume that typically varies from 1.1 to 1.7, and M_f is the free-stream

Mach number. The convective heat transfer coefficient H_c could be determined from the mathematical analyses of fluid flow or from an empirical formula involving dimensionless numbers such as the Nusselt number (N_o), the Reynolds number (Re), and the Prandtl number (P_r). Typical values of these numbers have been specified under Section 2.3. For specific applications, relevant equations and data can be found in numerous heat transfer textbooks prescribed for graduate courses.

2.5.3 RADIATION MODE OF HEAT TRANSFER

The heat transfer through the radiation mode is very complex and, hence, is difficult to estimate or compute with high accuracy because of the large number of variables involved with a wide range of values. The heat transfer rate through radiation between two surfaces is given by

$$[dQ/dt] = [\sigma A F_e F_c (T_2^4 - T_1^4)] \text{W/sec} \quad (2.24)$$

where σ is the Stefan-Boltzmann constant (W/cm²/K), F_e is the emissivity factor, and F_c is the configuration factor dependent on geometry of the object and other variables involved. The emissivity factor can be replaced by the effective emissivity parameter ε_{eff}, which is defined as

$$\varepsilon_{\text{eff}} = [\varepsilon_1 \varepsilon_2 / \varepsilon_2 + \varepsilon_1 - \varepsilon_1 \varepsilon_2][1/N+1] \quad (2.25)$$

where ε_1 and ε_2 are the emissivities of the layers on opposite sides of the surface perpendicular to the direction of heat transfer, and N is the number of isolated surfaces or layers with various emissivities. The magnitude of effective emissivity for a double-aluminized Mylar layer varies from 0.008 for N equals 2 to 0.001 for N equals 10 and when both emissivities are equal to 0.05, the high temperature is 300 K, the low temperature is 77 K, and the pressure is 10^{-5} torr [4]. Equation (2.25) is valid when no significant heat leakage occurs through the blanket edges. In actual practice, the effective emissivity can be significantly higher because residual pressure may exist between the layers. Emissivity of the aluminized sheets may increase due to damage or contamination. In the (presiding equation, the impact of temperature is not clearly visible; nevertheless, a significant change in the hot-boundary temperature will change the true value of the effective emissivity and effective thermal conductivity. In summary, effective emissivity or effective thermal conductivity is extremely difficult to determine because of their dependency on several temperature-dependent factors.

2.6 Summary

This chapter briefly summarized the thermodynamic factors affecting the design of a heat exchanger in a cryogenic cooling system. Early developments in the refrigeration fields were briefly explained, and symbols and units of potential thermodynamic variables were provided. Partial derivatives of thermodynamic variables were summarized for when a specific variable or parameter is held constant. Laminar and turbulent heat flows that affect the heat exchanger performance and cooling system efficiency were discussed with the use of complex mathematical expressions. Frictional coefficient ratios (FCRs) for both laminar and turbulent flow environments were computed as a function of the through Reynolds number and the rotational Reynolds number. The ratio of the Nusselt number under rotating and static conditions was calculated and plotted as a function of the through Reynolds number and the rotational Reynolds number. The static Nusselt number was calculated as a function of the through Reynolds number and the Prandtl number. The two-fluid model, which can predict the heat exchanger performance with great accuracy, was described in greater detail with emphasis on normal-fluid and superfluid components. The two velocity components were derived in terms of density, entropy, temperature, complex variable, and Del operator. Convectional acceleration and viscous effects caused by the thermomechanical force were described with appropriate numerical examples. Mathematical expressions for three modes of heat transfer — conduction, convection, and radiation — were derived. Numerical examples for the computation of heat exchanger performance parameters were provided wherever possible. Thermodynamic formulas to obtain thermal analysis and computer modeling data on heat exchangers for integration of cryocoolers and microcoolers were derived [5].

References

1. J. R. Howell. *Fundamentals of engineering thermodynamics*. New York: McGraw-Hill, 1987, pp. 521–523.

2. W. D. Morris. *Heat transfer and fluid flow in rotating coolant chambers*. New York: Research Study Press, a division of John Wiley & Sons, 1981, pp. 158, 163.
3. T. Kitanura et al. "A numerical model on transient, two-dimensional flow and heat transfer in He II." *Cryogenics*, 1997, vol. 37, no. 1, pp. 2–4.
4. G. Zissis and W. Wolfe. *The Infrared Handbook*. 1978, pp. 15–69.
5. R. C. Weast (Ed.). *Handbook of chemistry and physics* (51st ed.). Boca Raton, FL: CRC Press, 1970–1971, p. F-96.

Chapter 3 | Thermodynamic Aspects and Heat Transfer Capabilities of Heat Exchangers for High-Capacity Coolers

3.0 Introduction

This chapter exclusively focuses on the thermodynamic aspects and heat transfer capabilities of the heat exchangers for high-capacity coolers. The most critical issues in the design of heat exchangers are addressed in great detail. Mathematical expressions for laminar and turbulent heat flows, heat transfer efficiency, overall heat transfer coefficients, and input power requirements are derived. Three distinct modes of heat transfer — conduction, radiation, and convection — are described with emphasis on heat balance and critical performance parameters that have a significant impact on the design of a heat exchanger.

3.1 Modes of Heat Transfer Phenomenon

Three distinct modes of heat transfer in solid, liquid, and gas media will be discussed in great detail. Mathematical equations identifying the heat flow from one medium to another as a function of heat-related variables are derived. Studies performed by the author indicate that the heat flow analysis under equilibrium thermodynamic conditions is not very complex compared to a heat flow analysis under transient environments. However, the laws of thermodynamics must be strictly observed irrespective of the operating environments. This means that the heat transfer phenomenon must obey the first and second laws of thermodynamics. The conservation of energy law states that the energy is conserved and heat must flow from a hot to a cold region. It is impossible for heat to be transferred from a cooler body to a hotter one. The principal objective of including the rudiments of heat-exchange thermal analysis is that heat exchangers are the essential elements of thermal systems, including conventional thermal and nuclear power plants, gas compressors, jet engines, steam turbines, condensers, refrigeration units, and high-capacity cooling systems.

Fundamental laws of heat transfer are well documented in several books on heat transfer theory. The laws of radiation and conduction of heat transfer are based on experimental investigations conducted by well-known scientists. The mathematical analysis for the convective heat flow is not considered as a well-established law of heat transfer but rather is treated as an empirical expression involving several variable parameters with a wide range of values.

3.2 Three Distinct Laws of Heat Transfer

The conduction mode of heat transfer indicates the heat transfer within the medium itself. In the case of metals or alloys, conduction is due to two factors: the thermal drift of free electrons and photon vibration. At lower temperatures, photon vibration is due to the vibration of the crystalline structure, resulting in a flow of heat energy within the medium. In the case of liquids, the mechanism for conduction is due to the combination of electron drift and molecular collision. Furthermore, conduction in liquids is strictly temperature dependent, not pressure dependent. In the case of gas, the mechanism for heat conduction is due to molecular collision and is dependent on both temperature and pressure.

Radiation is considered to be a thermal energy flow between two bodies separated by a distance. Radiation is based on the electromagnetic waves. The radiated heat flow is proportional to the fourth power of the surface temperature, but it is directly proportional to the surface area of the body radiating the heat energy.

The convective heat transfer occurs between a solid surface and cooling fluid. This is a mixed mode of heat transfer, in which the heat at the solid-fluid interface is transferred by the conduction mode involving molecular collisions between the solid and the fluid molecules. Due to these collisions, a temperature change occurs in the fluid, a density variation is produced, and a bulk fluid motion occurs. In a mixed mode, there is a mixing of the high- and low-temperature fluid elements leading to a heat transfer between the solid and the fluid through convection [1]. In the case of fluids, properties of fluid dynamics are of critical importance when addressing heat transfer issues.

3.3 Description of Heat Transfer Modes

3.3.1 CONDUCTION

The law of conduction for heat transfer states that the conductive heat flow (Q_c) is proportional to the thermal conductivity of the material (k), the surface area (A) normal to the heat flow direction, and the temperature difference between the hot and cold surfaces, but it is inversely proportional to the separation (L) between the two surfaces operating at temperatures T_h (hot surface temperature) and T_c (cold surface temperature). Thus, the expression for the heat flow can be written as

$$Q_c = [kA(T_h - T_c)]/L \quad \text{BTU/hr or W} \tag{3.1}$$

Computed values of heat flow as a function of area (A), length, or separation (L) and temperature difference (dT) between hot and cold surfaces are summarized in Table 3.1.

These computations assume a thermal conductivity (k) of 0.5 W/m·K and indicate that the conductive heat transfer increases with the increase in surface area, thermal conductivity, and temperature differential. However, the conductive heat transfer decreases with the increase in wall thickness.

Table 3.1 Computed Values of Heat Flow for Various Wall Thicknesses (L) and Areas (Watt).

	L = 20 cm		L = 40 cm	
DT (K)	A = 0.1 m²	A = 1 m²	A = 0.1 m²	A = 1 m²
50	12.5	125	6.25	62
100	25	250	12.5	125
200	50	500	25	250
250	62.5	625	31.25	312.5
400	100	1000	50	500

Thermal resistance is similar to electrical resistance and can be defined as

$$R_t = [L/kA] \qquad (3.2)$$

where L is the separation between two surfaces, k is the thermal conductivity of the wall, and A is the surface area. Inserting this equation into Equation (3.1), one gets the heat flow expression as

$$Q_c = [T_h - T_c]/R_t \qquad (3.3)$$

In this particular case, the parameter R_t is called conductive thermal resistance. It is important to mention that in the case of steady-state flows, the thermal conductivity (k) does not vary with temperature; however, under turbulent flow environments, its magnitude undergoes radical change, particularly at higher operating temperatures.

3.3.2 RADIATION

The law for radiation heat transfer was first discovered by two distinguished scientists, J. Stefan through experimental investigations and L. Boltzmann using statistical methods [1]. This law states that the radiant heat flow (Q_r) for a black body is proportional to the surface area (A) times the fourth power of temperature. Thus, the black body emissive power (P_e) can be written as

$$P_e = [\sigma A T^4] \text{W} \qquad (3.4)$$

where $\sigma = 5.67 \times 10^{-8}$ W/m²·K⁴, T = black body temperature (K), and A = area (m²).

Note that a black body absorbs the entire radiation incident upon it. Equation (3.4) does not indicate the net radiative heat transfer (Q_r) between the two equal surface areas with operating temperatures of T_2 and T_1. The net heat transfer between the two surfaces through radiation can be expressed as

$$Q_r = [\sigma A (T^4 - T^4)] \text{W} \qquad (3.5)$$

Since the real body surface areas are not perfect absorbers and radiators, they emit a fraction of the black body radiation for the same surface temperatures. This fraction is called the emittance-radiated power and can be written as

Table 3.2 Radiative Heat Transfer between a Gray Body and Room (W).

Surface area A (m²)	T_2/T_1	$\varepsilon = 0.5$		$\varepsilon = 0.06$	
		1000/400 K	1500/500 K	1000/400 K	1500/500 K
0.05		1381	7087	1657	8505
0.10		2762	14,174	3314	17,010
0.20		5524	28,348	6628	34,020
0.40		11,048	56,696	13,256	68,040

$$Q_r = [\sigma A \varepsilon (T_2^4 - T_1^4)] \, \text{W} \tag{3.6}$$

where ε is the emissivity of the radiating surface that is dependent on surface temperature. One can compute the radiative heat transfer between the gray surface body and room environments as a function of gray surface emissivity. Computed values of radiated heat transfer as a function of emissivity are shown in Table 3.2. These computations reveal that heat transfer through radiation increases rapidly as the temperature difference between the surface and the room increases. Note that surface conditions are included in the emissivity parameter that can reduce the gray body emission signatures as well as the radiated heat transfer capability.

3.3.3 CONVECTION

The heat transfer through convection is dependent on conduction heat transfer, fluid flow type (laminar or turbulent), and the mixing process. It is equally important to note that the convection mode of heat transfer is of paramount importance in the design of heat exchangers. There are two types of convection modes: free convection mode, where the density changes cause the bulk fluid motion, and the forced convection mode, where a pressure differential causes the bulk fluid motion, which is normally created by a mechanical pump or fan. The convective heat transfer or flux can be empirically defined as

$$Q_{\text{conv}} = [AC_c(T_s - T_f)] \quad \text{BTU/hr or W} \tag{3.7}$$

where A is the surface area (m²), C_c is the coefficient of convection (W/m²·K), T_s is the solid surface temperature (K), and T_f is the tempera-

Table 3.3 Typical Magnitudes of Convection Coefficient for Various Flow Modes.

Fluid flow mode	C_c (W/m²·K)
Free convection in air	5–20
Forced convection in air	20–200
Free convection in water	20–100
Forced convection in water	50–10,000

ture of the fluid far away from solid surface (K). One watt is equal to 3.413 BTU/hr.

The preceding expression in terms of heat resistance can be written as

$$Q_{conv}[T_s - T_f]/R_{conv} \quad \text{BTU/hr or W} \tag{3.8}$$

where

$$R_{conv} = [1/C_c A] \tag{3.9}$$

The magnitude of the coefficient of convection depends on physical parameters of the surface, chemical properties of fluid such as viscosity, and fluid velocity over the solid surface area (A). These parameters influence the rate at which thermal energy enters or leaves the fluid medium. The impact of various flow modes on the coefficient of convection is evident from the data summarized in Table 3.3.

The value of the coefficient under forced convection in water is dependent on the velocity and temperature profiles of the cold fluid moving over a heated surface, as illustrated in Figure 3.1. It is evident from the velocity and temperature profile curves that the velocity component decreases as it approaches the solid surface or reaches in the fluid layer next to the surface. The heat transfer from the solid surface to the fluid occurs through conduction because the fluid layer has zero velocity, which is equal to heat transfer by convection into the rest of the flow. This means that

$$Q_{conv} = [k_f A][\partial T/\partial y]^w_{y=0} \tag{3.10}$$

where k_f is the thermal conductivity of the fluid, A is the surface area, T is the temperature, and w is the wall dimension. It is evident from this expression that the heat flux to surface area ratio is directly proportional to the thermal conductivity of the fluid and the temperature gradient at the

3. Thermodynamic Aspects for High-Capacity Coolers

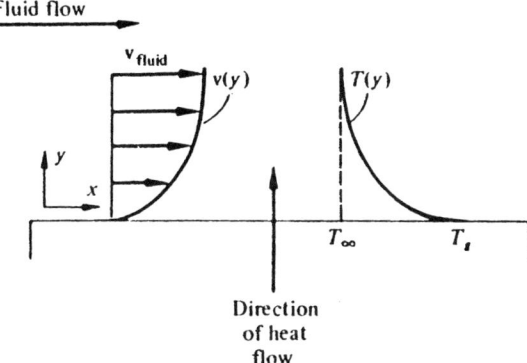

Fig. 3.1 Velocity and temperature profiles of a cold fluid moving over a hot surface. Symbols: $v(y)$ = amplitude of the velocity profile along y-axis; $T(y)$ = amplitude of the temperature profile along y-axis; T_s = temperature at the body surface (K°); and T_∞ = temperature at far distance from the surface (K°). *Note*: The direction of the heat flow is along the y-axis, whereas the temperature is along the x-axis.

wall. It is equally evident that the temperature is directly proportional to the fluid velocity with high velocities permitting higher temperature gradients. In brief, convection heat transfer in a tube is dependent on the fluid velocity, fluid properties such as density and viscosity as a function of temperature, and the overall convection heat transfer coefficient. Note that the fluid viscosity varies with the temperature radically, particularly, at higher temperatures. The fluid properties can be expressed in dimensionless parameters such as flow through the Reynolds number (*Re*), the Prandtl number (*Pr*), and the Nusselt number (*Nu*).

The flow through Reynolds number is defined as

$$Re = [vD/\mu_k] \qquad (3.11)$$

where v is the fluid velocity, D is the internal diameter of the tube, and μk is the fluid kinetic viscosity, which is the ratio of (μ/e), where μ is the fluid dynamic viscosity. For Reynolds numbers ranging from 2300 to 6000, the laminar flow begins a transition to turbulent flow in a tube. However, the flow is completely turbulent at Reynolds numbers exceeding 6000. Furthermore, the turbulent flow has a flatter velocity profile than laminar flow because of the pronounced effect of fluid velocity on the convection heat transfer coefficient. This is an important aspect of fluid dynamics.

The dimensionless Prandtl number can be expressed as

$$Pr = [\mu C_{p/k}] \tag{3.12}$$

C_p is the specific heat of the fluid flowing, and all other variables have been defined in previous equations.

The Nusselt number depends on flow regimes such as laminar flow and turbulent flow involving liquids or gases, viscosity as a function of temperature and density, tube physical parameters, fluid temperature, the Reynolds number, and the Prandtl number. For a laminar flow [1] in a short tube, the most reliable expression for the Nusselt number can be written as

$$Nu = \left[(1.86) Re^{0.33} Pr^{0.33} (D/L)^{0.33} (\mu/\mu_w)^{0.14}\right] \tag{3.13}$$

where D is the internal pipe diameter, L is the pipe length, and μ_w is the fluid dynamic viscosity near the wall surface that varies with temperature.

In the case of turbulent flow, where the temperature difference between the tube surface and average fluid temperature is less than 5°C for liquids or 55°C for gases, the Nusselt number can be given as

$$Nu = \left[(0.023) Re^{0.8} Pr^n\right] \tag{3.14}$$

where n has a value of 0.4 for a heating and 0.3 for a cooling mode.

In cases where the temperature limits exceed or the fluid viscosity is greater than that of water, the expression for the Nusselt number can be written as

$$Nu = \left[(0.027) Re^{0.8} Pr^{0.33} (\mu/\mu_w)^{0.14}\right] \tag{3.15}$$

In Equation (3.15), all parameters are expressed at the average fluid temperature except the dynamic fluid viscosity at the wall, which is given at the tube wall temperature. It important to mention that all fluids have a definite resistance to change of form. This property is characterized as an internal friction known as viscosity and is expressed in dynes-second/cm² or poises. Note that the kinetic viscosity is the ratio of the absolute viscosity to the density of the fluid. The absolute viscosity of water at 20°C is 0.01002 poise. The c.g.s unit for the kinetic viscosity is stoke. Conversion factors [2] are summarized in Table 3.4. The absolute viscosity (poise or gr/cm·sec) is equal to the product of kinetic viscosity (stoke) and density (gr/cm³). Densities of various fluids as a function of temperature

Table 3.4 Viscocity Conversion Factors.

	Viscosity conversion *Klnematic*	
To convert from	**To**	**Multiply by**
cm^2/sec (Stokes)	Centistokes	10^2
	ft^2/hr	3.875
	ft^2/sec	1.076×10^{-2}
	$in.^2/sec$	1.550×10^{-1}
	m^2/hr	3.600×10^{-1}
$cm^2/sec \times 10^2$	cm^2/sec (Stokes)	1×10^{-2}
(Centistokes)	ft^2/hr	3.875×10^{-2}
	ft^2/sec	1.076×10^{-5}
	$in.^2/sec$	1.550×10^{-3}
	m^2/hr	3.600×10^{-3}
ft^2/hr	cm^2/sec (Stokes)	2.581×10^{-1}
	$cm/sec \times 10^2$ (centistokes)	2.581×10
	ft^2/sec	2.778×10^{-4}
	$in.^2/sec$	4.00×10^{-2}
	m^2/hr	9.290×10^{-2}
ft^2/sec	cm^2/sec (Stokes)	9.29×10^2
	$cm^2/sec \times 10^2$ (centistokes)	9.29×10^4
	ft^2/hr	3.60×10^3
	$in.^2/sec$	1.44×10^2
	m^2/hr	3.345×10^2
$in.^2/sec$	cm^2/sec (Stokes)	6.452
	$cm^2/sec \times 10^2$ (centistokes)	6.452×10^2
	ft^2/hr	2.50×10
	ft^2/sec	6.944×10^{-2}
	m^2/hr	2.323
m^2/hr	cm^2/sec (Stokes)	2.778
	$cm^2/sec \times 10^2$ (centistokes)	2.778×10^2
	ft^2/hr	1.076×10
	ft^2/sec	2.990×10^{-3}
	$in.^2/sec$	4.306×10^{-1}

continued

Table 3.4 *Continued*

Viscosity conversion
Absolute
Absolute viscosity = kinematic viscosity × density; lb = mass pounds;
lb_F = force pounds

To convert from	To	Multiply by
gm/(cm)(sec) [Poise]	gm/(cm)(sec)(10^2) [Centipoise]	10^2
	kg/(m)(hr)	3.6×10^2
	lb/(ft)(sec)	6.72×10^{-2}
	lb/(ft)(hr)	2.419×10^3
	lb/(in.)(sec)	5.6×10^{-3}
	$(gm_F)(sec)/cm^2$	1.02×10^{-3}
	$(lb_F)(sec)/in.^2$ [Reyn]	1.45×10^{-5}
	$(lb_F)(sec)/ft^2$	2.089×10^{-3}
gm/(cm)(sec)(10^2) [Centipoise]	gm/(cm)(sec) [Poise]	10^{-2}
	kg/(m)(hr)	3.6
	lb/(ft)(sec)	6.72×10^{-4}
	lb/(ft)(hr)	2.419
	lb/(in.)(sec)	5.60×10^{-5}
	$(gm_F)(sec)/cm^2$	1.02×10^{-6}
	$(lb_F)(sec)/in.^2$ [Reyn]	1.45×10^{-7}
	$(lb_F)(sec)/ft^2$	2.089×10^{-5}
kg/(m)(hr)	gm/(cm)(sec)	2.778×10^{-3}
	gm/(cm)(sec)(10^2) [Centipoise]	2.778×10^{-1}
	lb/(ft)(sec)	1.867×10^{-4}
	lb/(ft)(hr)	6.720×10^{-1}
	lb/(in.)(sec)	1.555×10^{-5}
	$(gm_F)(sec)/cm^2$	2.833×10^{-4}
kg/(m)(hr) (Cont.)	$(lb_F)(sec)/in.^2$ [Reyn]	4.029×10^{-5}
	$(lb_F)(sec)/ft^3$	5.801×10^{-5}
lb/(ft)(sec)	gm/(cm)(sec) [Poise]	1.488×10^1
	gm/(cm)(sec)(10^2) [Centipoise]	1.488×10^3
	kg/(m)(hr)	5.357×10^3
	lb/(ft)(hr)	3.60×10^3
	lb/(in.)(sec)	8.333×10^{-2}
	$(gm_F)(sec)/cm^2$	1.518×10^{-2}
	$(lb_F)(sec)/in.^2$ [Reyn]	2.158×10^{-4}
	$(lb_F)(sec)/ft^2$	3.108×10^{-2}

3. Thermodynamic Aspects for High-Capacity Coolers

Table 3.4 *Continued*

To convert from	To	Multiply by
lb/(ft)(hr)	gm/(cm)(sec) [Poise]	4.134×10^{-3}
	gm/(cm)(sec)(10^2) [Centipoise]	4.134×10^{-1}
	kg/(m)(hr)	1.488
	lb/(ft)(sec)	2.778×10^{-4}
	lb/(in.)(sec)	2.315×10^{-5}
	(gm_F)(sec)/cm^2	4.215×10^{-6}
	(lb_F)(sec)/in.^{2} [Reyn]	5.996×10^{-5}
	(lb_F)(sec)/ft^2	8.634×10^{-6}
lb/(in.)(sec)	gm/(cm)(sec) [Poise]	1.786×10^{3}
	gm/(cm)(sec)(10^2) [Centipoise]	1.786×10^{4}
	kg/(m)(hr)	6.429×10^{4}
	lb/(ft)(sec)	1.20×10
	lb/(ft)(hr)	4.32×10^{4}
	(gm_F)(sec)/cm^2	1.821×10^{-1}
	(lb_F)(sec)/in.^{2} [Reyn]	2.590×10^{-3}
	(lb_F)(sec)/ft^2	3.73×10^{-1}
(gm_F)(sec)/cm^3	gm/(cm)(sec)	9.807×10^{3}
	gm/(cm)(sec)(10^2) [Centipoise]	9.807×10^{4}
	kg/(m)(hr)	3.530×10^{5}
	lb/(ft)(sec)	6.590×10
	lb/(ft)(hr)	2.372×10^{5}
	lb(in.)(sec)	5.492
	(lb_F)(sec)/in.^{2} [Reyn]	1.422×10^{-2}
	(lb_F)(sec)/ft^2	2.048
(lb_F)(sec)/in.^{2} [Reyn]	gm/(cm)(sec) [Poise]	6.895×10^{4}
	gm/(cm)(sec)(10^2) [Centipoise]	6.895×10^{6}
	kg/(m)(hr)	2.482×10^{7}
	lb/(ft)(sec)	4.633×10^{3}
	lb/(ft)(hr)	1.668×10^{7}
	lb/(in.)(sec)	3.861×10^{2}
	(gm_F)(sec)/cm^2	7.031×10
	(lb_F)(sec)/ft^2	1.440×10^{2}
(lb_F)(sec)/ft^2	gm/(cm)(sec) [Poise]	4.788×10^{2}
	gm/(cm)(sec)(10^2) [Centipoise]	4.788×10^{4}
	kg/(m)(hr)	1.724×10^{5}
	lb/(ft)(sec)	3.217×10
	lb/(ft)(hr)	1.158×10^{5}
	lb/(in.)(sec)	2.681
	(gm_F)(sec)/cm^2	4.882×10^{-1}
	(lb_F)(sec)/in.^{2} [Reyn]	6.944×10^{-3}

Table 3.5 Density of Various Fluids at Room Temperature.

Fluid	Temperature (°C)	Density (gr/cm³)
Acetone	20	0.792
Alcohol	20	0.791
Gasoline	20	0.680
Water	20	0.9982
Water	30	0.9956

Table 3.6 Viscosity of Liquids as a Function of Temperature [3].

Liquid	Temperature (°C)	Absolute viscosity (centipoise)
Air liquid	−192 (81 K)	0.172
Alcohol	20/30/40	1.362/1.070/0.914
Olive oil	20/40/70	84/36/12
Water	30/40/50	0.792/0.652/0.545

Table 3.7 Viscosity of Gases as a Function of Temperature [3].

Gas	Temperature (°C)	Absolute viscosity (micropoise)
Neon	20/100	311/365
Krypton	0/15	233/246
Nitrogen	11/27	171/178
Oxygen	20	202
Xenon	20	226

are shown in Table 3.5. The viscosity of liquids and gases as a function of temperature are summarized in Tables 3.6 and 3.7, respectively.

The specific heat (C_p) of water as a function of temperature is shown in Table 3.8.

Values of the variables in Table 3.8 as a function of temperature have been provided because these variables are required to compute the Prandtl numbers under fluid dynamic conditions. As stated earlier, the heat flux is dependent on the specific heat, surface area, and temperature difference between the two surfaces. In the case of water-cooling systems or steam power plant condensers, heat removal is strictly dependent on the thermal

Table 3.8 Specific Heat of Water as a Function of Temperature.

Temperature		Specific heat (J/gr·C)
°C	K	
20	293	4.183
30	303	4.178
40	313	4.179
50	323	4.181

Note: 1 Joule is equal to 4.186 Cal, and 1 gr.Cal is equal to 4.186 Joules or watt.sec.

Table 3.9 Thermal Conductivity (K) of Coolant as a Function of Temperature.

Temperature		Thermal conductivity (w/m·K)		
°C	K	Water	Helium	Nitrogen
0	273	0.5550	0.1405	0.0237
27	300	0.6091	0.1497	0.0267
77	350	0.6682	0.1649	0.0294
127	400	0.6861	0.1795	0.0325

Note: The thermal conductivity of water at 20°C or 293 K is 0.597 W/m.K.

conductivity of the cooling medium as a function of temperature (see Table 3.9).

Once the values of the variables, the coolant flow rate, and the pipe dimensional parameters are known, one computes the coefficient of convection heat transfer, the Reynolds number, the Prandtl number, and the Nusselt number.

Numerical Example

Compute the average value of the convection coefficient, the Reynolds number to determine the flow regime, the Prandtl number, and the Nusselt number for the 20°C-water flow entering into a 3-cm diameter steel pipe at velocities of 10, 20, 30, 40, and 50 m/sec.

Solution

Inserting the given parameters in Equation (3.11), one gets the Reynolds number value as

$$Re = [(5000)(3)]/[(0.01002)/(0.9980)] = 1.494 \times 10^6$$

Because of such a high value of Re, the flow is turbulent as illustrated in Figure 3.2. Reynolds numbers as a function of fluid velocity and pipe internal diameter are shown in Figure 3.2. It is evident from the curves shown that the Reynolds number exceeds 200,000 as soon as the flow velocity approaches 5 m/sec regardless of pipe diameter.

Inserting the given parameters in Equation (3.12), one gets the value of the Prandtl number as

$$Pr = [0.01002 \times 4.183/0.597] = 7.02$$

Inserting the values of Re and Pr numbers into Equation (3.13), one gets the Nusselt number as

$$Nu = \left[(0.0230)(1.48 \times 10^6)^{0.8}(7.02)^{0.4}\right] = [4363]$$

The convection coefficient can now be calculated using the following expression:

$$C_{conv} = [Nu\,k/D] \quad (3.16)$$

Inserting the given parameters for the 50 m/sec flow in Equation (3.16), one gets

$$C_{conv} = [4363 \times 0.597/0.03] = [86{,}824]$$

Note that all of these calculations have been performed for the flow rate of 50 m/sec and an internal pipe diameter of 3 cm. Similar computations are performed for the remaining flow rates and for pipe diameters of 1 and 3 cm. Computed values of Re, Pr, Nu, and C_{conv} as a function of flow rates and pipe internal diameters are shown in Table 3.10.

Computed values of the Nusselt number as a function of the Reynolds number (not exceeding 10,000) and the Prandtl number (not exceeding 4) are depicted in Figure 3.3. It is important to mention that according to Equation (3.12) and data presented in Table 3.9, the Prandtl number is independent of the fluid velocity and pipe diameter. It is further evident

Table 3.10 Computed Values of the Reynolds Number, Prandtl Number, Nusselt Number, and Coefficient of Convection Heat Transfer.

Fluid flow velocity (m/s)	Re	Pr	Nu	C_{conv} (W/m²·K)
1	29,880	7.02	191	3796
10	298,800	7.02	1204	23,959
20	597,600	7.02	2096	41,710
30	896,000	7.02	2898	57,670
40	1,245,000	7.02	3649	72,615
50	1,494,000	7.02	4363	86,824

from Table 3.9 that the flow is turbulent because all of the Reynolds numbers shown are greater than the assumed threshold of 6000. The tabulated data indicate that the coefficient of convection is directly proportional to the Nusselt number but inversely proportional to the pipe diameter.

3.4 Impact of Heat Transfer Modes on Heat Exchanger Performance

Most engineering applications involve a combination of two or three heat transfer modes. In the case of the composite wall shown in Figure 3.4, heat is transferred by the fluid at a higher temperature to the wall through convection and radiation modes of heat transfer. Elements of the thermal circuit for the planar wall (A) and composite (B) wall are shown in Figure 3.4.

3.4.1 HEAT TRANSFER IN A PLANAR WALL

In the case of the planar wall, heat within the wall medium is transferred by conduction. However, outside of the wall surface, heat is transferred through convection to the surrounding fluid at a relatively lower temperature. But at lower temperatures, radiation does not significantly contribute to heat transfer. Thus, on the furnace side or hot side of the wall, the total heat flow expression can be written as

$$Q_t = [Q_r + Q_{ch}] = [T_h - T_{wh}] \qquad (3.17)$$

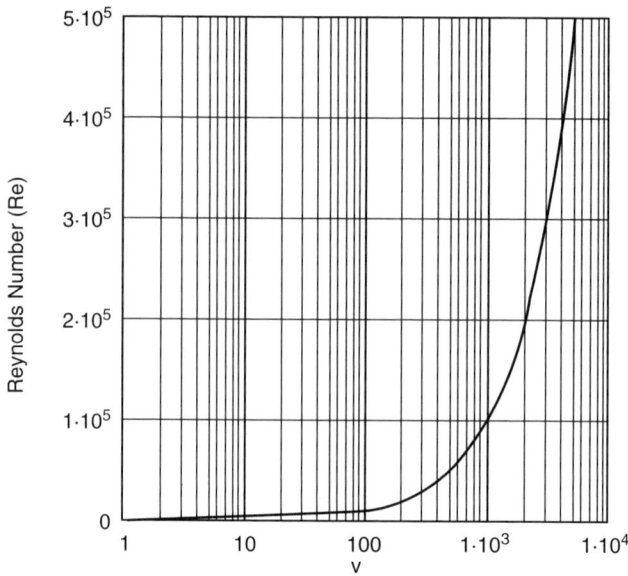
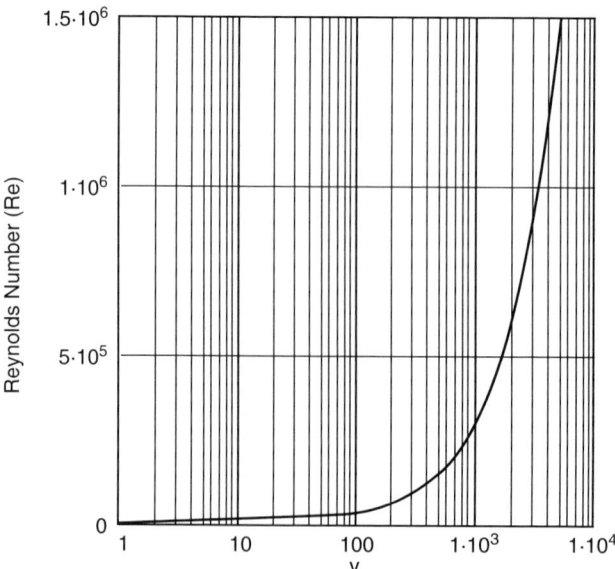

Fig. 3.2 Reynolds number as a function of fluid velocity and pipe diameter at room temperature.

3. Thermodynamic Aspects for High-Capacity Coolers 73

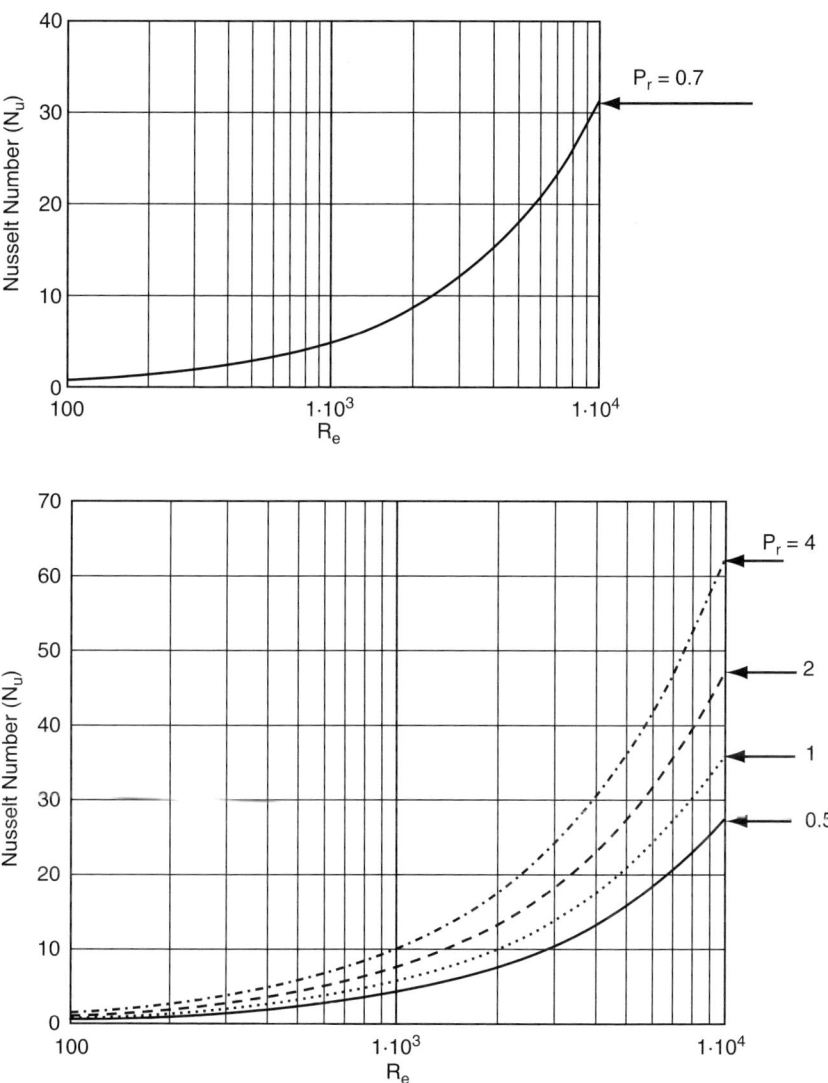

Fig. 3.3 Nusselt number as a function of Reynolds number (R_e) and Prandtl number (P_r).

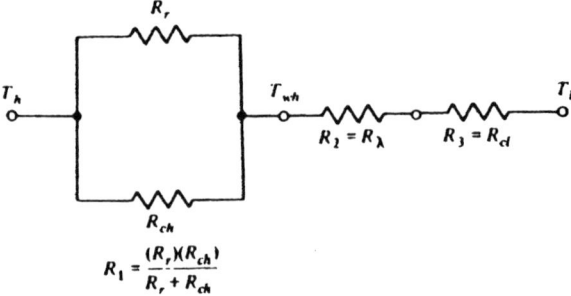

(A) Thermal circuit for planar wall

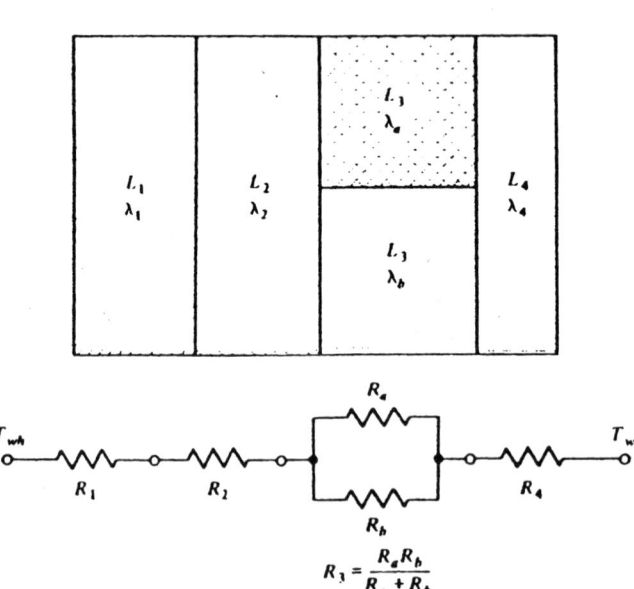

(B) Thermal circuit for composite wall

Fig. 3.4 Elements of the thermal circuit for the composite wall through which heat transfer has been considered. Symbols: T_h = high temperature, T_l = low temperature, T_{wh} = high wall temperature; T_{wl} = low wall temperature, R_1, R_2, R_3, and R_4 are thermal resistances; and L_1, L_2, L_3, and L_4 are the wall lengths with specified thermal conductivities (λ).

where Q_r is the heat transfer through radiation, Q_{ch} is the heat transfer from the hot side of the wall t through convection, T_h is the temperature of hot side, and T_{wh} is the temperature of the hot side of the wall. The thermal resistance of the first element of the planar wall can be expressed as

$$R_1 = [(R_{ch})(R_r)]/[R_{ch} + R_r] \quad (3.18)$$

where R_1 is the sum of the two thermal resistances that are connected in parallel as illustrated in Figure 3.4(A). However, the total heat flow by conduction can be written as

$$Q_t = [T_{wh} - T_{wl}]/R_2 \quad (3.19)$$

where R_2 is the thermal resistance of the region shown in Figure 3.4, subscript (wh) indicates the high temperature side of the wall, and subscript (wl) indicates the low temperature side of the wall.

At the outside surface of the composite wall, the total heat flow can be written as

$$Q_t = [T_{wl} - T_l]/R_3 \quad (3.20)$$

where R_3 is the thermal resistance of the third section of the wall. Multiplying each of these three equations by their respective thermal resistances and adding, one gets the expression for the overall heat flow for the planar wall (Fig. 3.4A) as

$$Q_{overall} = [T_h - T_l]/[R_1 + R_2 + R_3] = \left[dT\big/\sum\nolimits^3 R_j\right] = [C_{over} dT] \quad (3.21)$$

where dT represents the temperature difference between the high and low temperatures of the wall, R_j indicates the thermal resistance of the jth section or region of the wall, and C_{over} is the overall coefficient of heat transfer. From this equation one can see that the overall coefficient of heat transfer is equal to the reciprocal of the sum of all thermal resistances.

3.4.2 HEAT TRANSFER IN A COMPOSITE WALL

In the case of a furnace or a heat exchanger composite wall, different materials with different thermal conductivities (k or λ) are involved, which are connected in series and in parallel as illustrated in Figure 3.4(B). A furnace wall normally uses various types of fire bricks followed by the steel castings to provide high mechanical strength at elevated temperatures. For such a furnace wall, the overall heat flow in terms of temperatures and thermal resistance can be written as

$$Q_{overall} = [T_{wh} - T_1] \Big/ \sum\nolimits^4 R_j \qquad (3.22)$$

where R_j represents the sum of all four thermal resistances as shown in Figure 3.4(B).

3.5 Heat Transfer through Heat Exchanger Pipes

A heat exchanger involves several tubes or pipes carrying cooling fluid to remove the heat from hot surfaces. In the case of pipes, the area perpendicular to heat flow varies with radial distance as illustrated in Figure 3.5, and the conductance expressions [3] must take into account the variable parameters involved. Heat flow expressions for a case of simple cylindrical pipe and an insulated pipe (Fig. 3.5) are derived.

3.5.1 HEAT FLOW IN A CYLINDRICAL PIPE

The heat flow in the radial direction follows a Fourier law, and its expressions can be written as

$$Q_{pipe} = [kA(dT/dr)] \qquad (3.23)$$

where k is the thermal conductivity for the pipe material, A is the pipe area normal to the direction of heat flow, and r is radius of pipe. This equation can be written in terms of pipe radii, the length of the pipe, and their respective temperature levels as

$$Q_{pipe} = \left[\int Q(k)(1/r)dr\right] = \left[\int (2\pi kl)dT\right] \qquad (3.24)$$

where

$$Q(k) = [(2\pi k)(T_1 - T_2)] / [\ln(r_2/r_1)] \qquad (3.25)$$

and T_1 is the temperature at a radial distance r_1, T_2 is the temperature at a radial distance r_2, and k is the thermal conductivity of the pipe material. Note that the integration with respect to (dr) is carried out over the two

3. Thermodynamic Aspects for High-Capacity Coolers

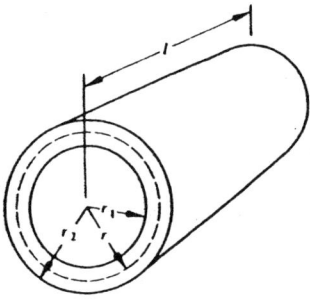

(A) Isometric view of a hollow cylinder

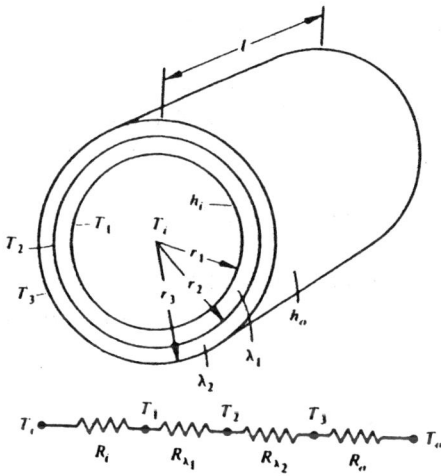

(B) Isometric view and thermal circuit for an insulated pipe

Fig. 3.5 Isometric views of a hollow cylinder and insulated pipe showing dimensions, surface temperatures, thermal conductivity of various sections, and elements of the thermal circuit. Symbols: T_o = output temperature; T_i = input temperature; l = length of pipe; h_i = heat intensity inside; and h_0 = heat intensity outside.

radial distances, while the integration with respect to temperature is carried out over the two temperatures, where T_1 is greater than T_2.

The heat flow in Equation (3.25) can also be written in terms of temperatures and thermal resistances as

$$Q(k) = [T_1 - T_2]/R(k) \qquad (3.26)$$

where the complex thermal resistance can be written as

$$R(k) = [\ln(r_2/r_1)]/[(2\pi kl)]. \qquad (3.26a)$$

It is interesting to mention that the inside temperature of the pipe is greater than the outside surface temperature, which means T_1 is greater than T_2. If the outside pipe temperature is greater than the inside pipe temperature, then the heat flow will be negative or in the reverse direction from outside to inside [3]. This situation occurs when the steel pipe is insulated (Fig. 3.5) to prevent heat loss or to prevent heat gain.

3.5.2 HEAT FLOW IN AN INSULATED PIPE

The geometry for the heat flow in an insulated pipe is illustrated in Figure 3.5. The heat flows from the center of the insulated pipe toward the pipe surface in the radial direction and its expression can be written as

$$Q_{in} = Q_1 = [(T_i - T_2)/R(\lambda_1)] \qquad (3.27)$$

where $R(\lambda_1)$ is the thermal resistance of Section 1, which is a function of thermal conductivity (λ or k) of the material used in Section 1 and is written as

$$R(\lambda_1) = [\ln(r_2/r_1)]/[(2\pi/\lambda_1)]$$

The heat flow for Section 2 with two temperature levels as shown in Figure 3.5 is given as

$$Q_2 = (T_2 - T_3)/R(\lambda_2) \qquad (3.28)$$

$R(\lambda_2)$ is the thermal resistance of region or layer 2 and is defined as $R(\lambda_2) = [\ln(r_3/r_2)]/[2\pi/\lambda_2]$, where λ_2 is the thermal conductivity of region or layer 2.

The expression for the heat flow from region 3 at temperature T_3 to outside surface temperature T_o can be written as

$$Q_o = (T_3 - T_o)/R_o \qquad (3.29)$$

where the outer surface thermal resistance $R_o = [1/2\pi r_3 h_o L]$. In this expression, parameter h_o represents the heat intensity at the outer surface of the pipe, and L is the length of the insulated pipe. The parameter T_i represents the temperature at the center of the pipe, T_1 at the inside surface of the pipe, T_2 at the interface of the pipe and insulation, T_3 at the insulated pipe external surface, and T_o as the outside ambient temperature. Furthermore, the temperature at the center of the pipe is at maximum, while the minimum temperature is far away from the insulated pipe's external surface. Note that thermal conductivity can be represented by either symbol (k) or (λ) while retaining the same unit of measurement (W/m·K).

3.6 Fundamental Design Aspects for a Heat Exchanger

It is of paramount importance to investigate and review the fundamental design aspects of a heat exchanger, because it is the most critical component of a cryogenic system. Design procedures must be adequately addressed prior to selection of a heat exchanger design configuration with particular emphasis on efficiency, reliability, safety, and cost-effectiveness. U-tube configuration, baffled-single-pass shell with fixed tubes, and shell-and-tube heat exchanger design with a floating head to accommodate differential thermal expansion between the tubes and shell are needed for high mechanical integrity. Potential heat exchanger design configurations shown in Figure 3.6 must be evaluated in terms of cost, power consumption, and efficiency. Once a heat exchanger design configuration is selected based on preliminary design analysis, its performance and physical parameters must be obtained through computer simulations. In addition, thermal performance level and pressure drops at specific points must be determined for the selected design candidate. Once these two steps are completed, the preliminary estimate for the size, weight, and reliability can be computed.

3.6.1 HEAT LOAD CALCULATIONS FOR HEAT EXCHANGERS

Heat load calculations for a specific heat exchanger configuration can be performed using the overall heat transfer coefficient (C_{over}) for the fluids involved such as water to water, water to condensing ammonia, water to lubricating oil, or ammonia to water. As stated earlier, the initial design requires calculations of thermal performance and pressure drops for both the cold and the hot stream or flow. Then an estimation of the individual heat transfer coefficient (C_{indu}) as a function of fouling factors is required.

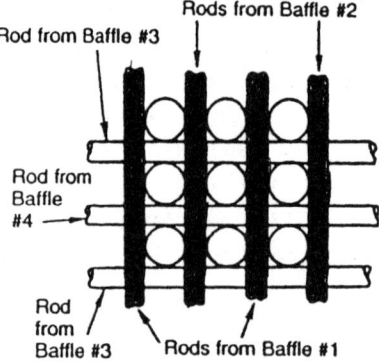

(A) Configuration with square-tube layout

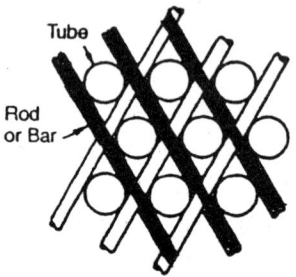

(B) Configuration with triangular-tube layout

Fig. 3.6 Most efficient and cost-effective layouts of tubes with rods. (A) Square layout, (B) triangular layout.

Typical heat transfer coefficients for the shell-and-tube type of heat exchangers are summarized in Table 3.11, whereas rough estimates of overall heat transfer coefficients for various fluids are shown in Table 3.12. The parameters shown in these tables will be most useful in the computation of the thermal performance parameters of a heat exchanger.

It is important to mention that the overall heat transfer coefficient depends on several parameters including the outside diameter of the tubes used in a specific heat exchanger layout (shown in Fig. 3.6), individual heat exchanger coefficients involving wall and thermal resistances under clean and fouling conditions, and the overall surface heat transfer efficiency (η_{over}). The heat load for a specific heat exchanger (Q_{ex}) configuration can be determined from the heat balance equation shown in Equation (3.30).

3. Thermodynamic Aspects for High-Capacity Coolers

Table 3.11 Typical Heat Transfer Coefficients for Shell-and-Tube Heat Exchangers.

Fluid type	Fluid condition	C_S $(W/m^2 \cdot K)$
Sensible heat transfer		
Water	Liquid	5000–7500
Ammonia	Liquid	6000–8000
Light organics	Liquid	1500–2000
Medium organics	Liquid	750–1500
Heavy organics	Liquid	
	Heating	250–750
	Cooling	150–400
Very heavy organics	Liquid	
	Heating	100–300
	Cooling	60–150
Gas	1–2 bar abs	80–125
Gas	10 bar abs	250–400
Gas	100 bar abs	500–800
Condensing heat transfer		
Steam, ammonia	No noncondensable	8000–12,000
Light organics	Pure component, 0.1 bar abs, no noncondensable	2000–5000
Light organics	0.1 bar, 4% noncondensable	750–1000
Medium organics	Pure or narrow condensing range, 1 bar abs	1500–4000
Heavy organics	Narrow condensing range, 1 bar abs	600–2000
Light multicomponent mixture, all condensable	Medium condensing range, 1 bar abs	1000–2500
Medium multicomponent mixture, all condensable	Medium condensing range, 1 bar abs	600–1500
Heavy multicomponent mixture, all condensable	Medium condensing range, 1 bar abs	300–600
Vaporizing heat transfer		
Water	Pressure <5 bar abs, $\Delta T = 25$ K	5000–10,000
Water	Pressure 5–100 bar abs, $\Delta T = 20$ K	4000–15,000
Ammonia	Pressure <30 bar abs, $\Delta T = 20$ K	3000–5000
Light organics	Pure component, pressure <30 bar abs, $\Delta T = 20$ K	2000–4000
Light organics	Narrow boiling range, pressure 20–150 bar abs, $\Delta T = 15$–20 K	750–3000
Medium organics	Narrow boiling range, pressure <20 bar abs, $\Delta T_{max} = 15$ K	600–2500
Heavy organics	Narrow boiling range, pressure <20 bar abs, $\Delta T_{max} = 15$ K	400–1500

Table 3.12 Rough Estimates of Overall Heat Transfer Coefficients (C_t) for Various Fluids for Possible Applications in Heat Exchanger Designs.

Fluids	C_t (W/m²·K)
Water to water	1300–2500
Ammonia to water	1000–2500
Gases to water	10–250
Water to compressed air	50–170
Water to lubricating oil	110–340
Light organics ($\mu < 5 \times 10^{-4}$ Ns/m²) to water	370–750
Medium organics ($5 \times 10^{-4} < \mu < 10 \times 10^{-4}$ Ns/m²) to water	240–650
Heavy organics ($\mu > 10 \times 10^{-4}$ Ns/m²) to lubricating oil	25–400
Steam to water	2200–3500
Steam to ammonia	1000–3400
Water to condensing ammonia	850–1500
Water to boiling Freon-12	280–1000
Steam to gases	25–240
Steam to light organics	490–1000
Steam to medium organics	250–500
Steam to heavy organics	30–300
Light organics to light organics	200–350
Medium organics to medium organics	100–300
Heavy organics to heavy organics	50–200
Light organics to heavy organics	50–200
Heavy organics to light organics	150–300
Crude oil to gas oil	130–320
Plate heat exchangers: water to water	3000–4000
Evaporators: steam/water	1500–6000
Evaporators: steam/other fluids	300–2000
Evaporators of refrigeration	300–1000
Condensers: steam/water	1000–4000
Condensers: steam/other fluids	300–1000
Gas boiler	10–50
Oil bath for heating	30–550

$$Q_{ex} = [(dm/dt\, c_p)_c (T_{c2} - T_{c1})] = [(dm/dt\, c_p)_h (T_{h1} - T_{h2})] \quad (3.30)$$

where dm/dt is the mass flow rate (F_r), T indicates temperature, subscript c stands for the cold stream, h stands for hot stream, c_p is the specific heat, and subscripts 2 and 1 for the cold stream indicate outlet and inlet temperature, respectively, while 1 and 2 for the hot stream indicate inlet and outlet temperature, respectively. Heat load calculations as a function of flow rates are shown in Figure 3.7.

Heat loads will be different for the two most common types of heat exchangers, which include a recuperative heat exchanger and a regenerator heat exchanger. In the case of a recuperative heat exchanger design, the hot fluid flow entering the system from the left is cooled by the cold fluid entering from the right as illustrated in Figure 3.8. However, in a regenerative system, the hot fluid entering the system from the left is cooled to a lower temperature by the cold matrix material. It is important to mention that the heat is first transferred from the hot stream at temperature T_h to the heat exchanger walls at temperature T_{wall} over the exchanger length. The heat transfer is proportional to the heat transfer coefficient between the wall and the hot stream fluid. Finally, the heat is transferred from the exchanger wall to the cold fluid, which is proportional to the heat transfer coefficient between the wall and cold fluid stream.

Numerical Example

Compute the hot stream inlet temperature for water entering the exchanger at 18°C and leaving at 42°C at a flow rate of 15,000 gm/sec and outlet temperature of 70°C for the hot stream.

Solution

The heat load for the exchanger with cold stream inlet and outlet water temperatures of 18°C and 42°C, respectively, is 1.506×10^6 W using the curve shown in Figure 3.7. Inserting the given parameters and computed heat load value in Equation (3.30), one gets

$$T_{h2} = [70 - (1.506 \times 10^6 / 4.184 \times 15,000)] = [70 - 24] = 46°C$$

3.6.2 COMPUTATION OF HEAT LOAD FOR THE HEAT EXCHANGER (Q_{ex})

It is evident from Equation (3.30) that the heat load for the cold fluid is dependent on mass flow rate, the specific heat of the cold water, and the

Heat Load Calculations for the Heat Exchanger $_uk$

$F_r := 5000, 10000 .. 50000$ gram per sec

$c_p := 4.184$ J per gr K $T_{c2} := 273 + 42$ K $T_{c1} := 273 + 18$ K

$Q_e(F_r) := F_r \cdot c_p \cdot (T_{c2} - T_{c1})$ W

$Q_e(F_r) =$

$5.021 \cdot 10^5$
$1.004 \cdot 10^6$
$1.506 \cdot 10^6$
$2.008 \cdot 10^6$
$2.51 \cdot 10^6$
$3.012 \cdot 10^6$
$3.515 \cdot 10^6$
$4.017 \cdot 10^6$
$4.519 \cdot 10^6$
$5.021 \cdot 10^6$

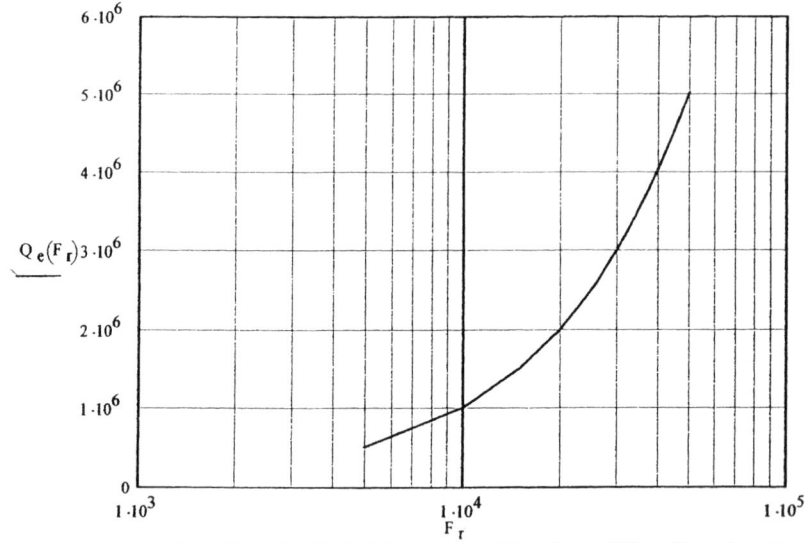

Heat Duty for Cold Stream as a Function of Flow Rate (watt)

Fig. 3.7 Heat load calculations for the cold stream in a heat exchanger.

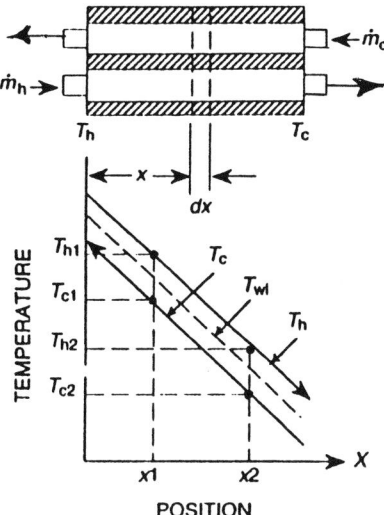

Fig. 3.8 Fluid temperature distributions in a recuperator heat exchanger. Symbols: T_c = temperature of the cold stream, T_h = temperature of hot stream; $\dot{m}_h$ = hot fluid flow rate, $\dot{m}_c$ = cold fluid flow rate; T_{wall} = temperature of exchanger wall; T_{c1} = cold stream input temperature, T_{c2} = cold stream output temperature; T_{h1} = hot stream output temperature, T_{h2} = hot stream input temperature; *Note:* In this particular heat exchanger configuration, the two fluids are continuously flowing through the heat exchanger. The temperature distribution curves shown here indicate the temperature difference between the two flows over the length of the heat exchanger, in which the hot fluid enters from the left and is cooled down by the cold fluid entering the exchanger from the right as shown in the figure.

input and output temperatures of the cold stream (Table 3.13). Specific heat calculations of water as a function of temperature are summarized in Table 3.14. Computed values of heat load as a function of flow rate and cold stream input and output temperatures are tabulated as follows.

3.7 Estimates of Heat Removal by Cold Water and Forced Air

Empirical formulas can be used to estimate the amount of heat removed through cold water cooling and forced air cooling. The following formulas are available for such estimates; nevertheless, the accuracy of the calculations is not better than 10%.

Table 3.13 Computed Values of Heat Load (Q_{ex}) for Cold Water Stream (kW).

Temperature (°C)		Flow rate (gm/sec)		
T_{c2}	T_{c1}	10,000	15,000	20,000
40	16	920	1381	1841
44	20	1004	1506	2008
50	25	1046	1569	2092

Table 3.14 Specific Heat of Water as a Function of Temperature.

	Specific heat	
Temperature (°C)	(cal/gr/°C)	(J/gr/°C)
0	1.0073	4.2177
1	1.0065	4.2143
20	0.9988	4.1817
30	0.9980	4.1783
50	0.9985	4.1803
100	1.0069	4.2163

$$H_{water} = [(266)(\text{gallons/sec})(T_0 - T_i)] \text{W} \quad (3.31)$$

$$H_{air} = [(169)(\text{cu.ft/sec})(T_0/T_i - 1)] \text{W} \quad (3.32)$$

where H is the amount of heat removed (watts), and T_0 and T_i are the outlet and inlet temperatures, respectively, of the cooling agent.

3.8 Computation of Overall Heat Transfer Coefficient, C_{oht}

The overall heat transfer coefficient (C_{oht}) for a heat exchanger [4] involves several variables and parameters including the shell-side heat transfer coefficient (C_s), the tube-side heat transfer coefficient (C_t), the tube internal radius (r_i), the thermal conductivity of the tube

Table 3.15 Thermal Conductivity of Metals at Various Temperatures (W/cm/K).

Metal	80 K	100 K	200 K	300 K	400 K
Aluminum	4.00	3.00	2.37	2.38	2.40
Copper	5.71	4.83	4.13	3.98	3.92
Gold	3.52	3.45	3.27	3.15	3.12
Iron	1.68	1.32	0.94	0.81	0.69

Table 3.16 Ambient Specific Heat for Certain Metals and Gases (cal/gr·C).

Metal/gas	Medium	Specific Heat (cal/gr·C)
Aluminum	Metal	0.226
Beryllium	Alloy	0.425
Copper	Metal	0.092
Gold	Metal	0.031
Iron	Metal	0.108
Magnesium	Alloy	0.249
Helium	Gas	1.242
Nitrogen	Gas	0.249

Note: To convert these values into J/gr·C units, simply multiply by a factor of 4.1868.

material (k), and the fouling resistance of the tube (R_{ft}). The thermal conductivity of various metals as a function of temperature is shown in Table 3.15.

It is important to point out that as the temperature increases, the thermal conductivity of the metals decreases. The thermal conductivity values can be converted into BTU/hr.ft.F units by multiplying the previous values by a factor of about 58. Thus, for iron its value will be about 47 in BTU/hr.ft.F at 300 K temperature. In some thermal calculations, the specific heat of metals and certain cooling gases may be involved. Values of specific heat(c_p) for some metals, alloys, and cryogenic gases at room temperature (20°C), as summarized in Table 3.16, can be used for thermal calculations.

3.8.1 COMPUTATION OF OVERALL HEAT TRANSFER COEFFICIENT UNDER FOULING CONDITIONS

It is important to mention that the reciprocal of the overall heat transfer coefficient is the sum of various heat transfer coefficients connected in parallel similar to various thermal resistances connected in parallel. The various heat transfer coefficients include the contribution from the shell-side, tube-side, environmental factor, and the metallic tube surfaces. Studies performed by the author indicate that the environmental factor has a significant impact on the magnitude of the overall heat transfer coefficient for the heat exchanger. The overall heat transfer coefficient [4] for the clean and fouling conditions will be computed as a function of various parameters involved.

3.8.1.1 Overall Heat Transfer Coefficient for the Foul Condition

First, the overall heat transfer coefficient for the fouling condition [4] will be computed using the following expression:

$$[1/C_{foul}] = [1/C_s + (D_o/D_i)(1/C_t)] + R_{ft} + (D_o/2)(1/k)Ln(D_o/D_i)] \quad (3.33)$$

Inserting the assumed values of 4500 W/m²·K (from Table 3.11) for C_s, 3500 W/m²K (from Table 3.12) for C_t, 175×10^{-6} m²·K/W (typical value for steel tubes) for the thermal resistance under fouling conditions (R_{ft}), 55 W/m·K (reasonable value) for the thermal conductivity (k) of steel, 0.020 meter (20 mm assumed) for outer tube diameter (D_o), and 0.016 meter (16 mm assumed) for the inner tube diameter (D_i) in Equation (3.33), one gets

$$[1/C_{foul}] = [1/4500 + (0.020/0.016)(1/3500) + 175 \times 10^{-6} + (0.020/2)(1/55)] \times Ln(0.020/0.016)]$$
$$= [.000222 + .000357 + .000175 + .000041] = [.000795]$$

or

$$[1/C_{foul}] = [0.000795]$$

This yields a value of 1258 W/m²·K for the overall heat transfer coefficient C_{oht} when the thermal conductivity (k) of steel is assumed as 55 W/m·K. The overall heat transfer coefficient values are 1239, 1251, and 1264 W/m²·K, when k has a value of 40, 40, and 60 W/m·K, respectively.

These computed values indicate that the impact of thermal conductivity on the overall heat transfer coefficient is at minimum or less than 1%.

3.8.1.2 Overall Heat Transfer Coefficient under Clean Environment

The heat transfer coefficients for the shell and tube will be different under clean environments. All contributing coefficients will be added in parallel as stated previously. The overall heat transfer coefficient under clean environments is given as

$$[1/C_{clean}] = [1/C_s + (D_o/D_i)(1/C_t) + (D_o/2)(1/k)Ln(D_o/D_i)] \quad (3.34)$$

Inserting the assumed value of 4800 for C_s (from Table 3.11), 5800 for C_t (from Table 3.12), 55 for thermal conductivity (k) for steel tubes, 0.024 for the outer diameter D_o, and 0.020 for the internal diameter D_i, one gets

$$[1/C_{clean}] = [1/4800 + (.024/.020)(1/5800) + (.024/2)(1/55)Ln(0.024/.020)]$$
$$= [.000208 + .000207 + .000040] = [0.000454]$$

or $C_{clean} = 2199 \, W/m^2 \cdot K$

These calculations for the overall heat transfer coefficient indicate that its magnitude under foiled conditions is about 57% of the value under clean environments assuming the same parametric values. The author has performed parametric analysis for the overall heat transfer coefficient assuming various parameters and the data shown in Table 3.17.

Computed values of the overall heat transfer coefficient indicate that its magnitude increases with the increase in thermal conductivity of the metallic tube and in fouling resistance. These computations further indicate that higher values of overall heat transfer are possible with larger tube diameters.

3.9 Computation of Critical Parameters for Heat Exchanger

Critical parameters of the heat exchanger — including the number of tubes and shells, tube layout constant, tube inner and outer diameters, shell diameter, temperature difference for the counter current flow, tube surface areas, tube cross-sectional areas under clean and fouling environments, and overall size of the system — need to be addressed to

Table 3.17 Parametric Analysis Data for Overall Heat Transfer Coefficient, C_{over} (W/m²·K).

K (W/m·K)	$C_s = 4500$, $C_t = 3500$ $D_o = .020$, $D_i = .016$ $R_{ft} = 175 \times 10^{-6}$	$C_s = 5000$, $C_t = 4000$ $D_o = .024$, $D_i = .020$ $R_{ft} = 200 \times 10^{-6}$	$C_s = 5000$, $C_t = 4000$ $D_o = .024$, $D_i = .020$ $R_{ft} = 150 \times 10^{-6}$
40	1235	1325	1419
50	1251	1344	1441
55	1258	1352	1450
60	1264	1358	1456

understand the overall performance capabilities and limitations of the heat exchangers.

3.9.1 COMPUTATION OF TEMPERATURE DIFFERENCE FOR THE COUNTERCURRENT FLOW

The temperature difference for the countercurrent flow can be computed from the given inlet and outlet temperatures. These temperatures include the cold fluid outlet temperature (T_{c2}), the cold fluid inlet temperature (T_{c1}), the hot fluid inlet temperature (T_{h1}), and the hot fluid outlet temperature (T_{h2}). The temperature difference [4] for the countercurrent flow can be determined from a heat balance expression, which is

$$[dT_{ccf}] = [(T_{h1} - T_{c2}) - (T_{h2} - T_{c1})]/[Ln\{(T_{h1} - T_{c2})/(T_{h2} - T_{c1})\}] \quad (3.35)$$

Assuming a cold fluid inlet temperature of 18°C, a cold fluid outlet temperature of 42°C, a hot fluid inlet temperature (T_{h1}) of 70°C, and a hot fluid outlet temperature (T_{h2}) of 34°C and inserting these values into Equation (3.35), one gets the differential temperature for the countercurrent flow as

$$[dT_{ccf}] = [(70 - 42) - (34 - 18)]/[Ln(28/16)]$$
$$= [28 - 16]/[0.5596] = [12/0.5596] = 21.4°C$$

However, the maximum temperature difference should not exceed 90% of this value. Thus, the maximum temperature difference $(dT)_{max}$ must not exceed 19.26°C for the preceding temperature values. Assuming a flow rate (dm/dt) of 10,000 gm/sec, heat load for the exchanger (Q_{ex}) of 1.506

TABLE 3.18 Computed Values of Hot Water Outlet Temperature Difference and Maximum Differential Temperature.

T_{h1} (assumed), °C	T_{h2} (calculated), °C	$(dT)_{ccf}$ (calculated), °C	$(dT)_{max}$ (calculated), °C
65	29	19	17
70	34	21	19
75	39	23	21
80	44	32	28

Note that these calculations are only valid for the assumed water inlet and outlet temperatures of 18°C and 42°C, respectively, a heat load of 1.506×10^6 W, and a water flow rate of 10,000 gr/sec. Any variation in these assumed variables will result in a change in the tabulated parameters.

$\times 10^6$ W, specific heat (c_p) of water of 4.184 W·sec/gr C, T_{c2} of 42°C, and T_{c1} of 18°C, and inserting these parametric values in Equations (3.30) and (3.35), one can compute the hot water outlet temperature (T_{h2}), differential temperature $(dT)_{ccf}$, and maximum temperature difference $(dT)_{max}$ as a function of several given or assumed parameters. The computed data for these three parameters are summarized in Table 3.18.

Note that these calculations are only valid for the assumed water inlet and outlet temperatures of 18°C and 42°C, respectively, a heat load of 1.506×10^6 W, and a water flow rate of 10,000 gr/sec. Any variation in these assumed variables will result in a change in the tabulated parameters.

3.9.2 COMPUTATION OF OUTSIDE SURFACE AREA (A_{os}) OF THE HEAT EXCHANGER

The physical size of a heat exchanger is dependent on the number of tubes, the outer diameter of the tube (D_o), the heat duty of the exchanger (Q_{ex}), the overall heat transfer coefficient (C_{over}), and the differential temperature function of four inlet and outlet temperatures for the cold and hot fluids. The outside heat transfer surface area can be written as

$$A_{os} = [\pi D_o L N_t] \qquad (3.36)$$

where L is the length of the tube, and N_t is the number of tubes in the heat exchanger. The diameter of the shell (D_s) is dependent on the number of tubes deployed and the diameter of individual tube. The total number of tubes can be computed by dividing the shell surface area by the net projected area of a single tube. This means that

$$N_t = [(C_{tc})(\pi D_s^2/4A_{os})] \tag{3.37}$$

where C_{tc} is the tube count constant that has a value of 0.95 for a one-tube pass configuration, 0.90 for a two-tube pass configuration, and 0.85 for a three-tube pass configuration due to the required clearance between the shell and the outer tube circle in case of a multiple-tube pass design. It is important to point out that the external heat transfer surface area is the A_{os} function of the outer diameter of the tube and heat load of the exchanger. Inserting Equation (3.36) into Equation (3.37), one gets

$$N_t = (0.785)[(C_{tc}/C_{tl})][D_s^2/(PR\,D_o)^2] = [A_{os}/\pi D_o L] \tag{3.38}$$

This equation yields the shell diameter as

$$D_s = [(0.6372)(PR)][(C_{tl}/C_{tc})(A_{os}D_o/L)]^{0.5} \tag{3.39}$$

The area of a single tube is dependent on the tube layout method, the pitch ratio (PR), and the outside tube diameter and can be expressed as

$$A_t = [(C_{tl})(PR\,D_o)^2] \tag{3.40}$$

where C_{tl} is the tube layout constant that depends on the layout geometry of the tubes.

3.9.3 ESTIMATION OF HEAT TRANSFER SURFACE AREA (A_{os}) UNDER CLEAN AND FOULING CONDITIONS

The heat transfer surface areas can be computed using the following expressions:

$$[A_{os}]_{clean} = [Q_{ex}/(C_{over})(dT)_{max}]_{clean} \tag{3.41}$$

$$[A_{os}]_{foul} = [Q_{ex}/(C_{over})(dT)_{max}]_{foul} \tag{3.42}$$

Inserting the computed values of overall heat transfer coefficients using Equations (3.32) and (3.33) for the foul and clean environments as 2199 and 1374 W/m²K, respectively, and assumed values of 1004×10^3 W for parameter Q_{ex} and $(273 + 19.2)$ K for parameter $(dT)_{max}$ in these two equations, one gets

$$[A_{os}]_{clean} = [1004 \times 10^3/2199 \times 292.2] = [1.56]m^2$$

$$[A_{os}]_{\text{foul}} = [1004 \times 10^3 / 1374 \times 292.2] = [2.50]\,\text{m}^2$$

These two values yield a ratio of outer heat transfer surface area under fouling and clean environments of 1.60.

3.9.4 COMPUTATION OF SHELL DIAMETER

Inserting the following assumed values of various constants and variables in Equation (3.39), one can compute the value of shell diameter. The assumed values for various parameters are:

$C_{tl} = 1.00$, $C_{tc} = 0.95$, $A_{os} = 2.50$
(as per pervious calculations under fouling condition),

Pitch ratio $(PR) = 1.2$, tube length $L = 2.00$, and diameter of tube $D_o = 20\,\text{mm}\,(0.020\,\text{m})$.

$$D_s = [(0.6372)(1.2)][(1.00/0.95)(2.50)(0.020)/(2.00)]^{0.5}$$
$$= [0.7646][0.1622] = [0.124\,\text{m}]$$

The shell diameter comes to 12.4 cm or 4.88 inches under the assumed parameters under the fouling conditions. However, the shell diameter is reduced to 3.86 inches, which is about 21% less than that under fouling conditions.

3.9.5 COMPUTATION OF NUMBER OF TUBES FOR THE HEAT EXCHANGER

Inserting the assumed and computed parameters shown under Subsection 3.9.2 in Equation (3.38), one can compute the number of tubes required for the heat exchanger as follows:

$$N_{\text{tube}} = [(0.7852)(0.95/1.00)][(0.124)/(1.2)(0.020)]^2$$
$$= [0.7459][26.69] = [19.9]$$

This means the number of tubes required using the assumed parameters and under fouling conditions is about 20, which appears low. However, if the PR and the outer tube diameter are reduced, the number of tubes required will increase.

These calculations can summarize the preliminary physical parameters for the heat exchanger for a given heat load and environmental condition. The parameters under fouling conditions are as follows:

Heat load:	1004×10^3 W (assumed)
Shell diameter:	0.124 meter (calculated)
Outside tube diameter:	0.020 meter (assumed)
Tube length:	2 meter (assumed)
PR:	1.2 (assumed)
Number of tubes:	20 (computed)

It is important to mention that the assumed parameters may not be accurate; however, they indicate a trend in the overall size of a heat exchanger. The preceding parameters appear to be reasonable for high-capacity cooling systems used in industrial applications.

3.10 Preliminary Rating of a Heat Exchanger

Once the preliminary parameters are obtained through calculations based on the assumed magnitudes of various constants and variables, these data can be fed into a computer program [4] to undertake a design analysis to meet the specified outlet and inlet flow temperatures and pressures.

3.11 Summary

This chapter has addressed heat transfer issues and thermodynamic aspects that are critically important in the design of heat exchangers for high-capacity cooling systems. Three distinct modes of heat transfer were discussed, giving practical examples for the conducting, convection, and radiation cases. Mathematical expressions were derived to identify the impact of various design parameters on the performance level of a heat exchanger for use in a high-capacity cryogenic system. Numerical examples have been provided to give readers a better understanding of critical design issues and design scenario for a heat exchanger. Expressions were derived for the Reynolds, Prandtl, and Nusselt numbers. The impact of Reynolds numbers, Prandtl numbers, and Nusselt numbers on fluid dynamics characteristics, turbulent flow, and laminar flow was discussed.

Specific heat, viscosity, thermal conductivity, and thermal resistance of various metals, gases, and water as a function of temperature were provided to assist in solving heat transfer problems. Heat balance and heat load computations as a function of mass flow, specific heat of the fluid and inlet and outlet fluid temperatures were provided for a rapid design analysis and to explain the importance of critical design issues. Design issues for recuperative and regenerator heat exchanger were briefly described with emphasis on heat transfer efficiency and reliability. Heat transfer coefficients were computed for the shell-side and tube-side design configurations as a function of thermal resistance under clean and fouling conditions, heat balance parameters, tube dimensions, and thermal conductivity of tube material. Numerical examples were provided under various subsections to compute shell diameter, the total number of tubes required to meet the desired performance level, the temperature difference for the counter current flow, and the overall size of the heat exchanger. Finally, critical parameters involved in the design of heat exchangers for high-capacity cooling systems were briefly summarized.

References

1. M. D. Burghardt. *Engineering thermodynamics applications* (3rd ed.). New York: Harper & Row, 1986, pp. 487–489.
2. R. C. Weast. *Handbook of chemistry and physics* (51st ed.). Boca Raton, FL: CRC Press, 1970–1971, p. F-35.
3. Burghardt. *Engineering thermodynamics applications*, pp. 496–497.
4. S. Kabac and H. Lic. *Heat exchangers*. Boca Raton, FL: CRC Press, 1998, pp. 268–269.

Chapter 4 | Critical Design Aspects and Performance Capabilities of Cryocoolers and Microcoolers with Low Cooling Capacities

4.0 Introduction

Cryocoolers and microcoolers have potential applications in high-resolution imaging sensors and high-performance missile systems, where a compact, low-cost cooling mechanism is of critical importance. In the case of military applications, cryocoolers or microcoolers are expected to provide improved reliability, high mechanical integrity, and enhanced performance under severe operating environments. In the case of microcoolers, maintaining a uniform temperature gradient and enhanced heat capacity are critical to achieve high heat exchanger efficiency with minimum power consumption. It is important to mention that in the case of space and airborne military systems, power consumption is of critical importance. Microcooler technology is best suited for space reconnaissance sensors, satellite communication equipment, air-to-air missiles, and high-resolution tactical systems where size, efficiency, reliability, and power consumption are of primary consideration. This chapter addresses critical design issues of cryocoolers and microcoolers, with an emphasis on efficiency, reliability, and power consumption. In addition to weight, reliability, efficiency, and power consumption, maintenance aspects for the cryocoolers will be addressed. Note that frequent maintenance and servicing of cryocoolers are required to ensure the safe, steady, and reliable operation of the cooling system.

4.1 Design Aspects and Operational Requirements

The following design aspects and operational requirements must be given serious consideration in the development of cryocoolers for specific space or military applications:

- Selection of a cryocooler must emphasize reliability and ease of operation with minimum cost and complexity.
- Integration of the cryogenic equipment with the superconducting device or system must be accomplished with minimum assembling cost.
- Field maintenance servicing requirements with minimum cost must be identified.

Compliance of these recommendations has been observed in relatively few commercial systems. However, the effective implementation of these recommendations has been observed in military infrared (IR) surveillance and target acquisition systems and space reconnaissance sensors to achieve specified mission objectives. Studies performed by the author indicate that the critical design aspects of both the closed-cycle and open-cycle cryocooler versions must be carefully evaluated with emphasis on cooling capacity, cost, and maintenance requirements. A review of currently operating refrigeration systems [1] and a survey of commercially available cryocoolers indicate that closed-cycle refrigerator systems have proven most cost-effective in terms of stability, reliability, and input power requirements. Typical input power requirements for a cryocooler as a function of heat load and operating temperature are shown in Figure 4.1. Reliable performance under specific operating conditions must be maintained with minimum cost and complexity, regardless of the cryocooler or microcooler type.

Tradeoff studies performed by the author reveal that Stirling cryocoolers are widely used in space and military IR sensors, whereas Collins and Gifford-McMahon (G-M) refrigerators are best suited for magnetically levitated trains, magnetic resonance imaging (MRI) equipment, and cryovacuum pumping systems as evident from Table 4.1. Both the high-temperature (77 K) and low-temperature (4.2 K) cryocoolers are clearly identified in Table 4.1 for specific applications [1]. The studies further indicate that the heat load capacity for a G-M cryocooler using Joule-Thomson (J-T) valve technology is about 100 mw at 4.2 K operation, but close to 1 W under higher cryogenic temperature (77 K) and controlled environments.

4.2 Performance Requirements of Cryocoolers

Operating temperature, cooling capacity, field maintenance, refrigeration efficiency, input power requirement, and reliability are the critical

4. Critical Design Aspects with Low Cooling Capacities

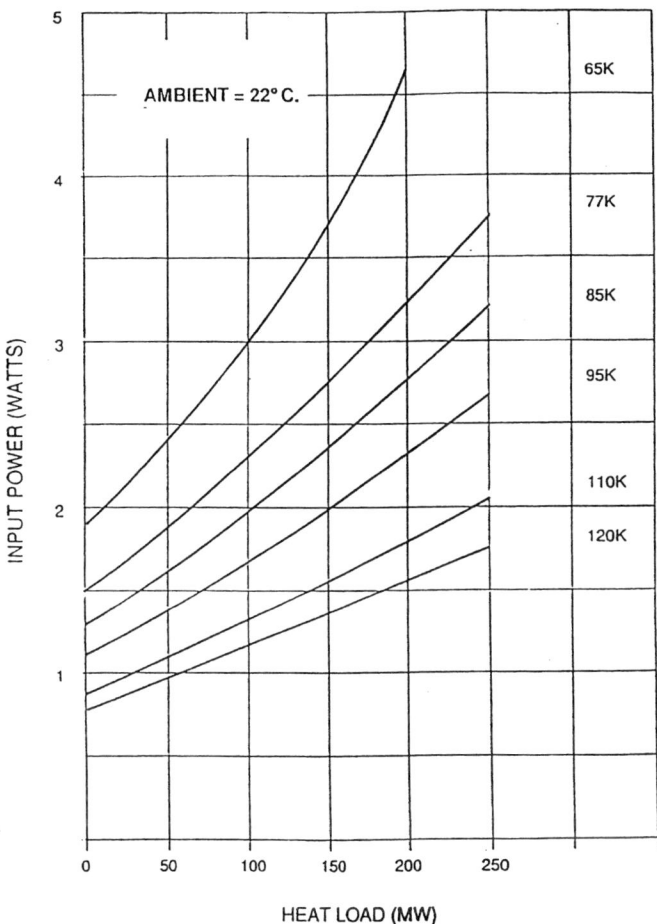

Fig. 4.1 Input power requirements for a cryocooler as a function of heat load and operating temperature.

operating and performance requirements of a cryocooler. Optimum performance with minimum cost and complexity should be the principal design criterion of any refrigerating system. In general, cooling capacity and weight of the cryocooler must be taken into account when comparing the overall performance levels of two cryocoolers. Theoretical analysis performed by the author on various cryocoolers indicate that the specific power requirements for a 5-watt cryocooler is about 15, 10, 5,

Table 4.1 Chronology of Systems Using Closed-Cycle Refrigeration.

Application	Temperature (K)	Refrigerator	User
1960s			
Infrared	80	Stirling	U.S. military
Hydrogen bubble chamber	20	GM	High-energy research
Microwave amplifiers	4.5	GM + JT	NASA
1970s			
High-energy accelerators	4.5	Collins	DOE
DC superconducting motor	4.5	Collins	U.S. Navy
Fusion	4.5	Collins	DOE
1980s			
Fusion magnet test facility	4.2	Collins	DOE
30-MJ SMES	4.5	Collins	BPA
Magnetohydrodynamics	4.5	Collins	Russia
Magnetically levitated train	4.5	Collins	Japan
Magnetic resonance imaging	4.2	GM	Commercial
Cryovacuum pumping	20	GM	Commercial
1990s			
SQUID	8.5	Stirling	NBS
SSC	2.0	Supercritical	DOE
MRI	10	GM	Commercial
Magnetically levitated train	—	—	DOT
SMES	—	—	Industry
Infrared	65	Stirling	U.S. military
Infrared	10	Stirling	NASA
High Tc	20–30	—	DOE/military

and 0.45 kW at cryogenic temperatures of 4.2 K, 10 K, 20 K, and 80 K, respectively. Whereas, the specific weight is about 200, 40, 15, and 1.5 kg/watt, corresponding to the previously stated cryogenic temperatures [2]. These parametric values indicate that significant reduction in input power level, weight, and size can be achieved at higher operating temperatures by performing parametric analysis. Tradeoff studies performed by the author indicate that significant improvement in input power requirements is possible simply by raising the operating temperature from 4.2 K

to about 8 K. The studies further indicate that conductive cooling is most desirable, because it is free from logistic and reliability problems [2].

4.2.1 MAINTENANCE ASPECTS AND RELIABILITY REQUIREMENTS FOR CRYOCOOLERS

Maintenance of a cryocooler or microcooler is critical if continuous and reliable performance over extended duration is the principal requirement. Maintenance cost and interval must be given serious consideration during the design phase of a cryocooler for a specific application. Limited maintenance published data available on the cryocoolers based on the field experience of the General Electric maintenance personnel [3] indicate that a cryocooler operating at a cryogenic temperature of 4.2 K requires more elaborate and frequent maintenance efforts than a cryocooler operating at a higher temperature (77 K). Higher maintenance costs are due to complex helium purification devices required to maintain clean working fluid at all times. Liquid helium cryocoolers normally run successfully and smoothly under controlled environments such as a research laboratory. However, when liquifiers are incorporated into a commercial application such as an MRI, the system fails to perform with high reliability. Refrigerator systems operating at higher superconducting temperatures have been more successful in meeting their stated reliability goals.

Cryocoolers used in space and military systems require stringent performance and reliability requirements because of harsh operating environments. Performance requirements for commercial and industrial refrigeration systems are somewhat relaxed due to higher cryogenic temperatures (Fig. 4.2). Three distinct commercial cryocoolers, namely dilution and magnetic refrigerators, Collins helium liquifiers, G-M cryocoolers, and G-M/J-T cryocoolers with Joule-Thomson valves, are widely used for low-temperature space and airborne missile systems operating over the 1 to 8 K temperature range.

Reliability of a high power, cryogenically cooled system is strictly dependent on the response and cooling capacity of a refrigeration system. A high-power system can be cooled with a Stirling-cycle cryocooler designed with a cooling capacity of 250 W at 77 K cryogenic temperature. This particular cryocooler is best suited for sonar transducers operating at cryogenic temperatures in the vicinity of 50 K with typical cooldown time not exceeding 4 hours. This cooler design is acoustically quiet and has demonstrated a mean-time-between-failure better than 50,000 hours. This

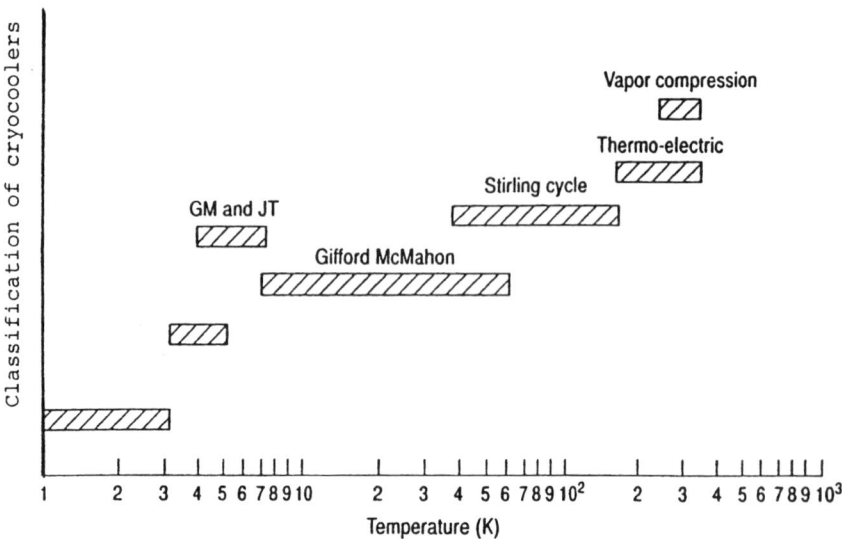

Fig. 4.2 Temperature ranges for commercial refrigerators.

cryocooler is most attractive for cooling submarine-based sonar transducers because of low acoustic signatures. The cooler can operate successfully in all directions and does not require the periodic replenishment of liquid nitrogen, and hence it is most suitable for ship propulsion superconducting systems. Closed-cycle coolers have been deployed in various commercial, space, and military applications because of improved reliability and cost-effective operation. Typical operating temperatures for commercial and military refrigerators are summarized in Figure 4.2.

4.2.2 COOLING POWER REQUIREMENTS FOR CRYOCOOLERS

Thermal analysis performed by the author indicates that the power required by the cryogenic refrigerators follows an exponential relationship to decreasing operating temperature. The analysis further indicates that by decreasing the operating temperature from 77 K to 50 K, the critical current in high temperature super conduction (HTSC) wires doubles, while the cooling power requirement increases by only 15%. A Stirling-cycle cryocooler designed with a typical capacity exceeding 250 watts at 77 K can efficiently cool a high-power system. Cooling power requirements depend on the operating temperature, the refrigerant used, and the overall thermal efficiency of the system.

4.2.2.1 Cooling Power Requirements for Microcoolers

Microcoolers have potential applications in space systems and sophisticated infrared military sensors including IR line scanners, IR search and track systems, thermal imaging radiometers, missile tracking systems, and airborne missile receivers. A microcooler operating on a Stirling-cycle engine principle can achieve the liquid helium temperature of 77 K in less than 3 minutes, which can be of critical importance in military applications. This particular microcooler configuration incorporates a regenerator to create a compression-expansion refrigerator cycle without valves. Such a regenerator has a large heat capacity and acts like an efficient heat exchanger. A microcooler developed for NASA incorporating a regenerator weighs only 15 ounces, consumes electrical power significantly less than 3 W, has a life expectancy of more than 5 years, and demonstrated a continuous operation exceeding 8000 hours without any failure. High reliability requires improved material technology with self-lubricating features, clearance seals with minimum friction, use of linear drives, and elimination of gaseous contaminants.

A microcooler configuration involving a split-Stirling cooler design with ratings from 0.25 W to 1 W and using linear drive can provide a continuous and reliable operation exceeding 7500 hours. Review of cryocooler technology reveals that a microcooler provides a cooling capacity or heat load of 150 mw at a temperature of 77 K with input power consumption less than 3.5 W and cooldown time ranging from 1.5 minutes to a cryogenic temperature 125 K to 4 minutes to a temperature of 77 K from a room temperature of 298 K. Input power requirements for various microcoolers are shown in Figure 4.1.

Microcoolers or crycoolers using rare-earth (RE) regenerator materials have demonstrated significant improvement in their performance levels. The first 4.2-K, two-stage G-M microcooler using erbium-nickel rare-earth regenerator technology achieved a temperature of 4.2 K in a remarkably short time because the regenerator material used has significant heat capacity below 10 K operating temperatures. This particular microcooler design using rare-earth material is of paramount importance in the IR seeker missiles and other battlefield sensors where time is the most critical element in tactical situations. The typical heat load capacity of such microcoolers is about 100 mw at 4.2 K operation; nevertheless, a heat capacity close to 1 W has been demonstrated at higher cryogenic operations under controlled environments and employing unique combinations of high-performance rare-earth materials.

4.3 Cryocoolers Using High-Pressure Ratios

Studies performed by the author on cryocoolers indicate that the use of a high-pressure expansion ratio and counterflow heat exchanger design are required to achieve a low-cost, high-efficiency cryocooler with enhanced reliability. The revolutionary design of the Boreas cryocooler shown in Figure 4.3 employs both the high-pressure ratio and counterflow heat

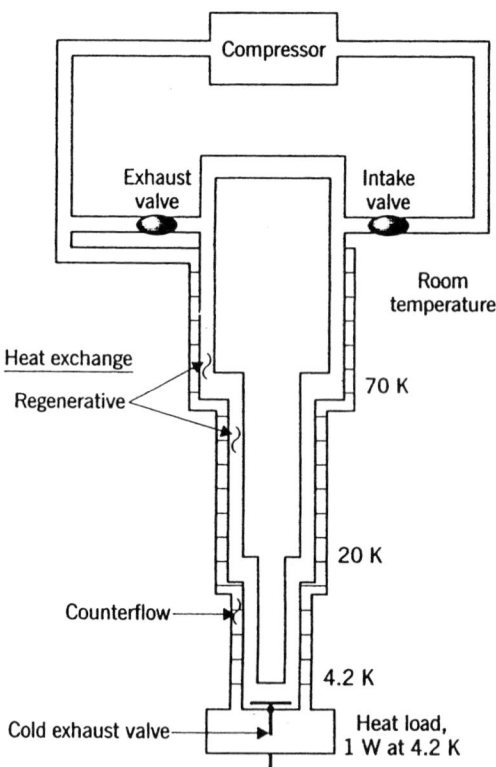

Fig. 4.3 Improved design of a three-stage Boreas cryocooler using advanced technologies.

4. Critical Design Aspects with Low Cooling Capacities

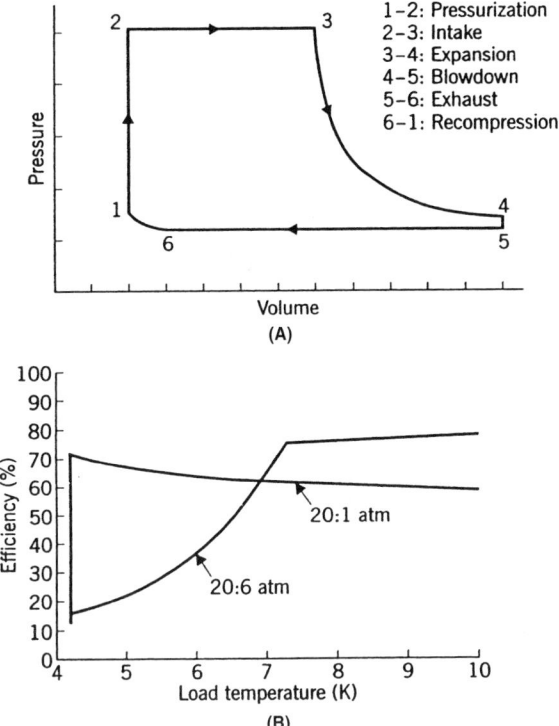

Fig. 4.4 (A) Pressure-volume diagram and (B) cold-stage efficiency of a three-stage Boreas cryocooler [4].

exchanger technology, thereby eliminating the need for costly high heat capacity rare-earth regenerator materials to achieve cost-effective cooling at 4.2 K operation. The three-stage Boreas cryocooler design [4] shown in Figure 4.3 provides a regenerator heat exchanger function in two warm stages at 70 K and 20 K temperatures, while the counterflow heat exchanger capability exists in the third stage to achieve a temperature of 4.2 K. It is important to mention that integration of the regenerator and the counterflow heat exchanger in a cooler design offers a compact and cold operation with a single helium flow circuit, thereby significantly reducing the cost and improving the reliability of the cryocooler.

The pressure-volume (P-V) diagram and the cold-stage efficiency as a function of load temperature (T_L) are shown in Figure 4.4 [4]. It is evident from the figure that the cold-stage efficiency improves with an increase in pressure. Note the (3–4) expansion cycle as illustrated in the figure reduces the gas temperature. Helium vaporized by the heat load is returned to the compressor via the counterflow heat exchanger, thereby recuperatively cooling the cylinder and the displacer.

4.3.1 ADVANTAGES OF HIGH-PRESSURE EXPANSION RATIO

The high-pressure expansion ratio significantly improves efficiency, minimizes power consumption, lowers operating costs, reduces helium mass flow, lowers cold-head speed, provides high reliability due to minimum wear and tear of parts, extends compressor life, and offers a vibration-free operation with a minimum of acoustic noise. A vibration-free cold head operation is highly desirable in applications, particularly where sensitive and delicate electronic components and circuits are used in the cooling system. Cryocoolers with high-pressure expansion ratios are best suited for MRI systems in hospitals, where patient comfort and minimum acoustic noise are the principal requirements.

Studies performed by the author on various cryocoolers indicate that the improved design of the Boreas cooler as illustrated in Figure 4.3 offers a quieter, vibration-free operation, thereby providing maximum comfort for the patients undergoing extended medical diagnostic tests conducted through high-resolution MRI systems. Cryocoolers operating with high-pressure expansion ratios open new opportunities for future applications that demand high efficiency, improved reliability, and virtually vibration-free continuous operation over extended periods.

4.4 Cooling Capacity of a Cryocooler

The maximum available cooling capacity (MACC) of a cryocooler is the most demanding performance parameter. Our discussion on the subject will be limited to a split-type free-displacer, the Stirling refrigerator shown in Figure 4.5, which has several advantages [5] including higher reliability and optimum cooling capacity. Thermal analysis performed by the author indicates that the net cooling capacity (Q_{net}) is dependent on the cold-head temperature (T_L), the normalized displacer natural frequency (f_n/n), and the displacer loss coefficient (C_d). The cooling capacity of this cryocooler as a function of these variables is shown in Figure 4.6. The MACC of this Stirling refrigerator system can be calculated by integrat-

4. Critical Design Aspects with Low Cooling Capacities

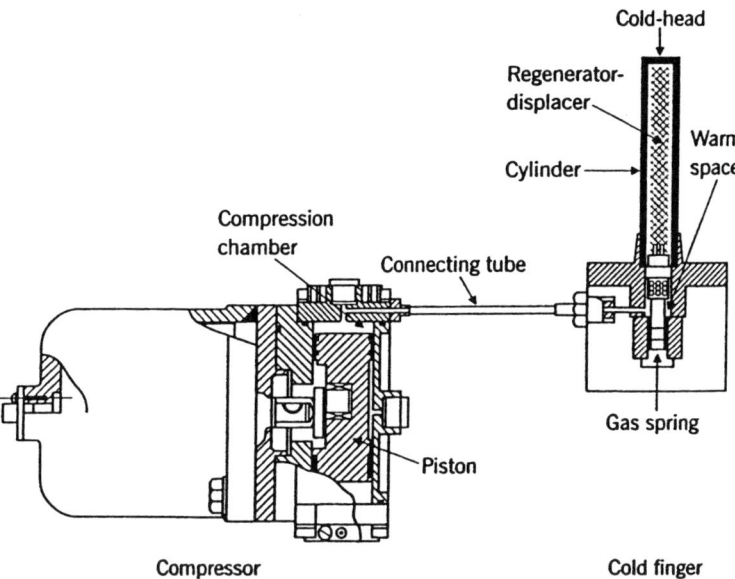

Fig. 4.5 Critical elements of a Stirling refrigerator design with optimum cooling capacity [5].

ing the expansion space represented by the pressure-volume curve shown in Figure 4.4. In brief, the quantity Q_{net} is proportional to the integral of expansion pressure and the displacer displacement as illustrated in Figure 4.4. In summary, the net cooling capacity is equal to the maximum available cooling capacity less various losses including the heat conduction loss in the regenerator, enthalpy flow loss in the regenerator, shuttle heat loss of the displacer, and hysteresis loss of the gas.

Operational analysis performed by the author indicates that the net cooling capacity can be obtained by the integral of the expansion pressure (P) and displacement volume (V). The net cooling capacity for a given compressor temperature (T_c), warm gas temperature (T_w), and displacer loss coefficient (C_d) increases as the cold-head temperature (T_c) increases. However, the net cooling capacity for a given cold-end temperature increases with the decrease of the displacer loss coefficient. It is interesting to point out that the maximum value of Q_{net} for given values of other temperatures involved occurs when the normalized displacer natural frequency approaches to 1.25, regardless of the displacer loss coefficient (C_{net}) as illustrated in Figure 4.6 [5].

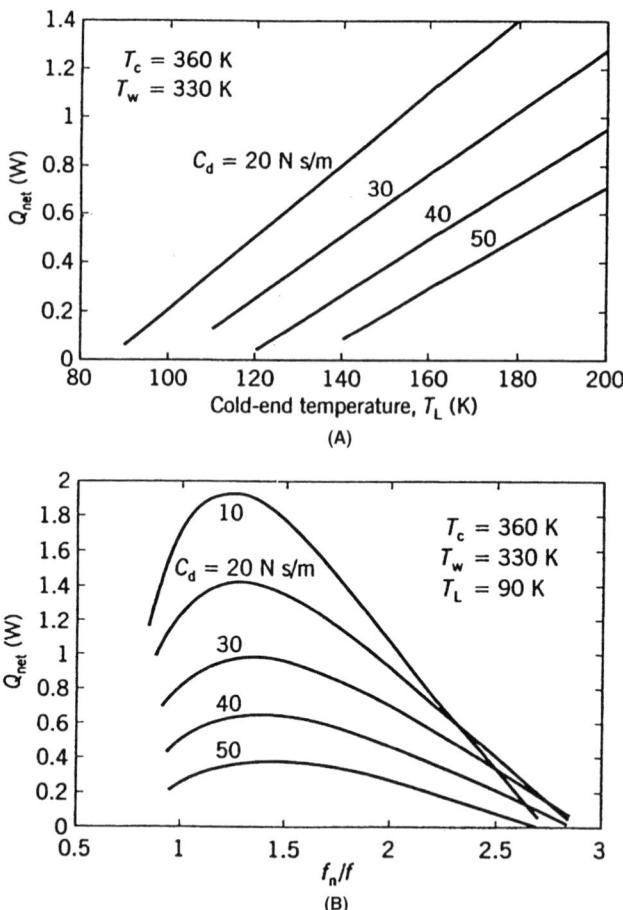

Fig. 4.6 Net cooling capacity (Q_{net}) (A) as a function of cold-end temperature and the displacer loss coefficient (C_d) and (B) net cooling capacity as a function of the normalized displacer natural frequency (f_n/f). Note: T_c and T_w indicate the temperature of the cold end and the warm end, respectively, in Kelvin.

The linear model used in the analysis indicates that for a given C_d, the net cooling capacity first increases with growing frequency, reaches a peak, and then decreases as shown in Figure 4.6. According to the curves shown in Figure 4.6, the optimum value of net cooling capacity occurs when the normalized natural frequency lies between 1.25 and 1.50. The magnitude of the displacer loss coefficient depends on the displacer seal design, the material used, and the quality of the workmanship provided in the cooler design. An accurate and reliable performance prediction of a cryocooler can be achieved through the modified linear network analysis. Furthermore, the displacer loss coefficient must take into account the frictional and gas leakage losses associated with the displacer. In summary, the net cooling capacity of a cryocooler depends on the cold-end temperature and the displacer loss coefficient that is a complex function of the displacer natural frequency (f_n) and the cold-head temperature (T_L).

4.5 Temperature Stabilization and Optimization of Mass Flow Rate

Reliable and efficient cryocooler performance requires temperature stabilization and optimization of mass flow rate. Transient behavior of a cryocooler is very important, particularly when the cryocooler is used in a sensitive space-based system or in a complex airborne weapon system. The transient behavior of a self-regulating cryocooler can be accurately predicted by the computer-based numerical analysis involving modeling of various bellow control mechanisms for a self-regulating function. A flow-regulating mechanism offers long-term, low gas consumption leading to maximum economy. A bellow control mechanism is generally used for gas flow regulation. The temperature-sensitive bellow senses the temperature at the cold end and then regulates the opening of the cooler orifice, thereby allowing adequate gas flow to maintain a specified cold temperature and to minimize the excess gas flow to avoid unnecessary gas wastage of refrigerant. Transient computer simulation requires various operating parameters, including mass flow rate (kg/sec), orifice opening (mm), nose-area ratio (unit-less), high-pressure temperature (K), low-pressure temperature (K), and mean temperature of the gas (K). The temperature-sensitive bellow mechanism allows a high flow rate to meet fast or slow cooling requirements and throttles down to conserve the cooling gas.

The regulation of the gas flow rate is contingent on the variation of the needle valve opening. The nose-area ratio is defined as the ratio of the area with the maximum valve opening to the area for a specific valve opening. The spatial temperature variation for the high-pressure gas and the glass Dewar depends on the needle-valve assembly position with respect to the initial position at various time intervals. Transient simulation indicates that it takes about 30 to 35 seconds for the spatial temperature to drop from 300 K to 80 K, depending on the environmental temperature.

4.6 Advanced Technologies for Integration in Cryocoolers

Research and development activities are currently directed toward the development of high-performance cryocoolers incorporating the latest technologies and advanced materials. Pulse tube refrigerators (PTRs) and advanced versions of thermal-electric (TE) coolers are currently getting great attention. Research and development studies performed by this author and other authors indicate that PTR technology has potential application in space and complex military airborne systems, where weight, size, power consumption, and reliability are of critical importance. The PTR cooling scheme shown in Figure 4.7 is the hottest new generation refrigeration technique that can be integrated into G-M and Sterling-cycle refrigerators. The latest research and development activities on PTRs indicate that the current unit cost varies between $15,000 and $20,000 depending on the cooling requirements and, thus, such cooling systems are not cost-effective at least at the present. However, the PTR system uses no moving parts, provides vibration-free operation, and offers ultra-high reliability. This particular refrigerator configuration is best suited for the most sensitive and complex sensors including IR missile receivers, electrooptic sensors for space and military applications superconducting quantum interference device (SQUID) sensors, cryopumps for high-density integrated circuits, and compact, high-resolution MRI equipment for medical diagnoses.

4.6.1 PULSE TUBE REFRIGERATION (PTR) SYSTEM DESIGN ASPECTS AND PERFORMANCE CAPABILITIES

In a PTR system (Fig. 4.7), oscillating pressures inside a pulse tube closed at one end will cause the gas to be heated at the closed end during the compression cycle and cooled at the other end during the expansion cycle. The compression and expansion of the gas are responsible for cooling as observed in other refrigerators. The heat exchange between the gas and pulse tube walls provides significant improvement in the conversion efficiency of the cooler [6]. Note that the absence of moving parts in the PTR systems offers the highest reliability, minimum noise, lowest maintenance

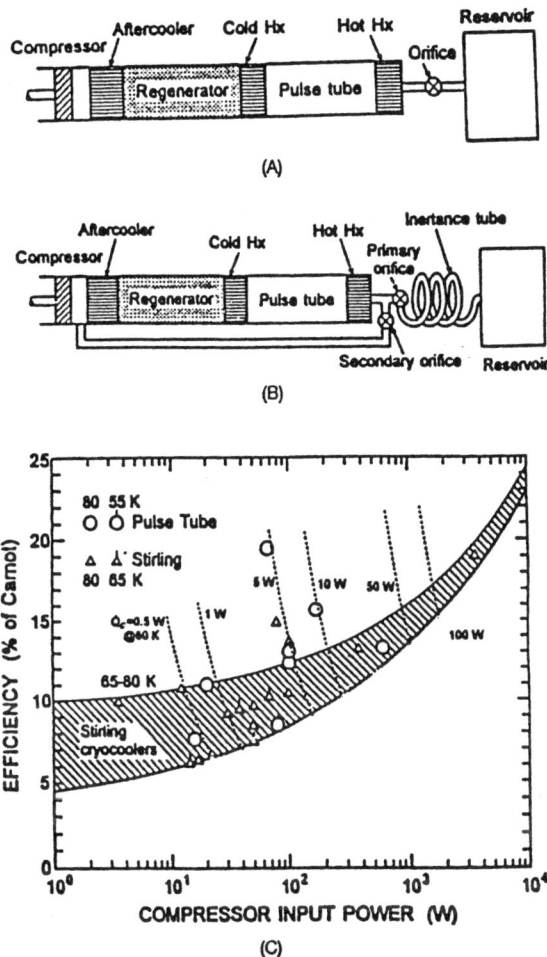

Fig. 4.7 Pulse tube refrigerator configuration incorporating (A) single orifice, (B) double-orifice scheme involving an inertance tube, and (C) efficiency of various coolers.

costs, and improved efficiency, unmatched so far by any other system. The addition of an orifice at the warm end of the pulse tube brings the temperature down to 105 K, resulting in an increase in the cooling power. The introduction of a second orifice (Fig. 4.7B) in the "double-inlet" design improves the efficiency very close to that of a Stirling-cycle refrigerator of comparable size. The double-inlet feature produces a heat lift of 30 W at 80 K operation with an efficiency of 13% of the Carnot cycle, which is considered the efficiency of an ideal refrigerator. The latest double-inlet PTR incorporating a critical component known as an inertance tube as shown in Figure 4.7(B) has demonstrated an efficiency close to 19% of Carnot's cycle, which is significantly higher than the best Stirling refrigerator efficiency. Performance comparison data including efficiency, compressor input power, and heat capacity for various refrigerator systems are shown in Figure 4.7(C).

To improve the performance of the PTR systems even further, select an optimum pulse tube configuration, proper tube orientation with respect to gravity, and helium gas as a working fluid. A thermodynamic model [6] developed for a single-PTR provides optimum performance for given system design parameters at a given temperature above 30 K and for a specified heat lift. The model indicates that a PTR system using optimum design parameters can achieve a cooling capacity exceeding 30 W at 60 K with a 1000-W compressor. Much lower cryogenic temperatures are possible with multistage PTR designs. A three-stage PTR developed by a Japanese professor in 1994 reached a cooling temperature of 4 K. A German scientist demonstrated a world's record cryogenic temperature of 2.2 K in 1997 using a two-stage PTR refrigerating system.

Research studies indicate that a recently developed mini-PTR system is capable of delivering 0.8 W at a cryogenic temperature of 80 K that has potential applications in complex military IR sensors, reconnaissance satellites, and high-resolution space systems. The smallest PTR [6] developed to date has a rating of 50 mw at 98 K, compressor swept volume of 0.75 cm^3, and a mechanical input power not exceeding 10 W. This particular most compact and efficient PTR system is best suited for space missions, space communications, and satellite-based reconnaissance. A PTR system requires a minimum of mechanical input power, thereby retaining the battery power level for extended periods, which is critical in spacecraft or satellites or in airborne reconnaissance sensors.

4.7 Classifications of Cryocoolers

Cryocoolers can be classified based on the operating cycle principle, namely, open-cycle and closed-cycle refrigerator systems. In certain applications involving low cryogenic temperatures, multistage refrigerators are required to meet the stringent cryogenic temperature requirements of superconducting systems such as MRI and sonar transmitter systems. Performance capabilities of both the high-temperature (77 K or higher) and low-temperature (4.2 K or lower) cryocoolers systems will be briefly discussed with emphasis on their unique features, such as cooling capacity, maintenance requirements, and efficiency. Cryocooler performance requirements for medical, scientific, and military applications are generally stringent, and these applications may require multistage-closed-cycle G-M or Stirling cryocoolers.

4.7.1 STIRLING-CYCLE CRYOCOOLERS

The latest research and development activities on bearing and seal technologies and the availability of high-quality lubricants have led to an optimum design of a Stirling cooler capable of demonstrating several years of continuous and unattended operation. Its potential applications include space sensors and unattended systems in remote or secluded areas where continuous operation with high reliability is the principal requirement. Stirling-cycle cryocoolers are widely used in applications where compact size and high reliability over extended periods are critical performance requirements. Such cryocoolers are best suited for cryogenic systems requiring low cooling capacity between 1 to 5 W over an operating temperature range of 60 to 80 K. These coolers use unique linear drive mechanisms instead of old rotary drive mechanisms and are in great demand for space and military applications where small weight and reliability are the critical performance requirements. The latest Stirling cooler design uses an orifice pulse tube, which eliminates the moving displacer from the expander, thereby leaving only one moving component in the compressor unit. Multistage designs of this particular cryocooler are currently in development and evaluation phases.

4.7.2 SELF-REGULATED JOULE-THOMSON (J-T) CRYOCOOLER

When Joule-Thomson valves are integrated with the Gifford-McMahon (G-M), liquid helium cooling is possible even below an operating temperature of 4.2 K with minimal loss of the refrigerant and without compromising the performance or reliability. The self-regulation of J-T valves further improves performance with no added complexity.

4.7.3 BOREAS-CYCLE CRYOCOOLER

Rigorous thermodynamic performance comparison between the G-M and Boreas-cycle cryocoolers operating at a cryogenic temperature of 4.2 K reveals that the Boreas-cycle cooler is four times more efficient than the G-M cryocooler. An improved design of a three-stage Boreas-cycle cryocooler incorporating advanced techniques and materials is shown in Figure 4.3. The performance comparison further reveals that the efficiency of a Boreas cycle is a strong function of the compressor suction pressure. It is evident from Figure 4.4 that the Boreas cycle offers higher efficiency than 70% with a suction pressure of 20:1 atm, while the G-M cycle typically operating with a suction pressure of 20.6 atm can achieve an efficiency of 17%. The higher efficiency of the Boreas cycle is due to the high-pressure expansion ratio. Furthermore, the high-pressure ratio is feasible only with a cryocooler employing a counterflow heat exchanger design concept.

4.7.4 CLOSED-CYCLE CRYOGENIC (CCC) REFRIGERATOR

The closed-cycle cryogenic (CCC) refrigerator, sometimes referred as a CCC cryocooler, is a viable alternative to a liquid helium cooling system. The CCC uses helium gas as the working fluid. The compact closed-cycle helium gas cryocooler shown in Figure 4.8 is best suited for superconducting quantum interference device (SQUID) sensor application because of its ultra-low temperature capability and high performance stability. This refrigerator has only two moving parts, a piston and a displacer, as depicted in Figure 4.8. Note that heat is produced during the compression mode and rejected during the expansion mode.

4.7.5 STIRLING CRYOCOOLER USING ADVANCED TECHNOLOGIES

Optimum design of Stirling cryocoolers has been possible due to active research and development activities in bearing and seal technologies and the successful development of high-quality lubrication. This particular cooler has demonstrated several years of continuous operation in remote locations without compromise in reliability and performance. These coolers are best suited for applications where size, weight, and reliability over extended periods are of critical importance. This cryo-

4. Critical Design Aspects with Low Cooling Capacities

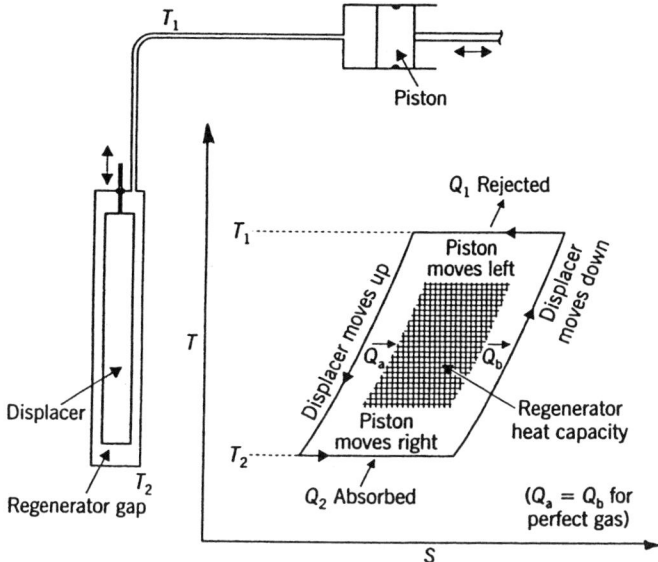

Fig. 4.8 Compact, closed-cycle, helium gas, split-type Stirling cryocooler.

cooler has a cooling capacity ranging from 1 to 5 W over a cryogenic temperature range of 55 to 80 K. Deployment of a unique linear drive mechanism in Stirling cryocoolers makes them most attractive for space and complex military applications because of high reliability over extended periods.

The single-stage Stirling cryocooler design [7] incorporating advanced nonmagnetic materials is shown in Figure 4.9. This cryocooler operates at 1 Hz and attains a temperature of 50 K in less than 4 hours that can be maintained for more than 2 days on a cylinder of helium gas. This kind of long duration operation without refill indicates the most efficient use of the helium by the cooling refrigerator. Based on the successful design of a single-stage Stirling cryocooler, a three-stage, closed-cycle, split-type Stirling cryocooler, shown in Figure 4.10, was developed using advanced materials and components: a nylon displacer, an epoxy-glass cylinder, and an aluminized-plastic radiation shield [7]. Note that the radiation shield eliminates the losses due to radiation, thereby ensuring high efficiency of the cryocooler. This particular cooler can maintain an operating temperature well below 15 K when operated at 1 Hz. It is important to mention that the elimination of the dead space has significantly improved the

116 Cryogenic Technology and Applications

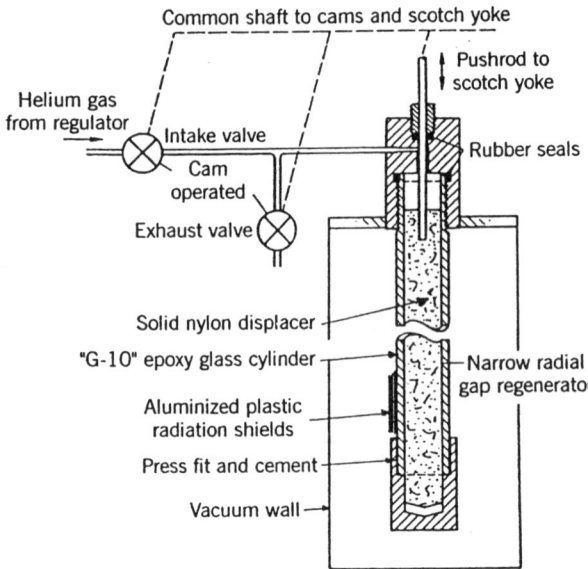

Fig. 4.9 Cross-sectional view of a single-stage, displacer-regenerator cryocooler using advanced nonmagnetic materials and narrow radial gap regenerator.

efficiency and reliability. This cooler uses a pulse tube design incorporating an orifice, which eliminates the moving displacer from the expander, thereby leaving only one moving component in the compressor unit. Multistage versions of this particular cooler with a wide range of cooling capacity are in the development phase and may open the door for future applications.

A microcooler design that incorporates advanced materials with an appropriate coefficient of expansion of various materials can provide effective optimization of the phase angle between the piston and the displacer. Optimization of this phase angle can maintain a cryogenic temperature below 13 K over a continuous duration as long as 5 weeks, which could be the longest operating period to date. This cryocooler has demonstrated a nonstop operation over 5000 hours with no sign of wear or tear in the components, while maintaining a temperature between 12.5 and 13 K at the cold end with no heat load. Two radiation shields are used: the outer radiation shield is maintained at 120 K, while the inner radiation shield is maintained at a temperature of 40 K. When operated at 1 Hz, the mechanical power required during the compression cycle is roughly

4. Critical Design Aspects with Low Cooling Capacities 117

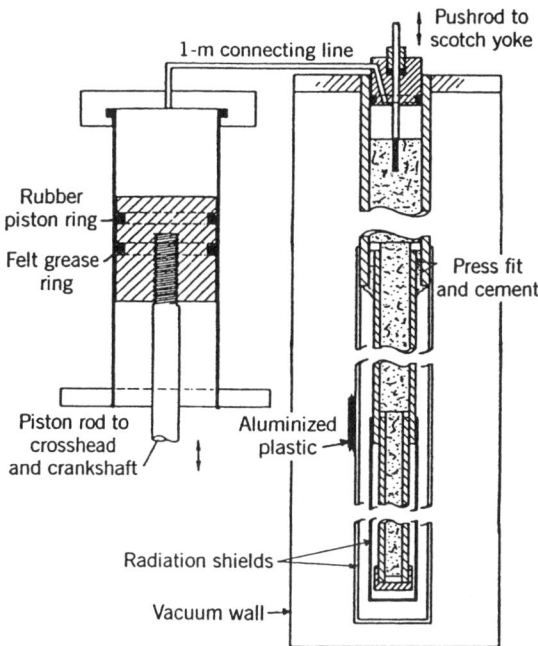

Fig. 4.10 Three-stage, closed-cycle, split-type Stirling cryocooler using advanced non-magnetic materials.

15 W, assuming an isothermal process. Since a consideration fraction of work is returned during the expansion cycle, the net mechanical power is hardly 10 W.

This particular cooler is capable of maintaining ultra-low temperatures over extended periods with very low power consumption. As stated earlier, this type of cryocooler can maintain operating temperatures well below 15 K with 10 to 15 mw of heat load. Note that a multistage cryocooler distributes the refrigerator capacity in such a way that the bulk of the heat input due to conduction and radiation is pumped at relatively higher temperatures, leaving little heat energy for lower temperatures. It is interesting to mention that a small liquid-helium cryostat with an evaporation rate of 1 liter per day of the helium gas at room temperature (300 K) can support a total heat load of about 30 mw at a cryogenic temperature of 4.2 and still have the capacity to absorb an additional load of 7.5 mw/K due to the heat capacity of the vapor. A market survey indicates that the cost of liquid helium will increase as the supply diminishes, whereas the cost of cryocoolers will decrease significantly if produced in large quantity.

4.7.6 G-M CRYOCOOLERS EMPLOYING J-T VALVES

G-M refrigerator configurations incorporating Joule-Thomson (J-T) valves are widely used in several commercial applications requiring cryogenic cooling at 4.2 K. It is interesting to note that from 1980 to 1995, both the superconducting MRI magnets and cryopumps made exclusive use of two-stage G-M cryocoolers to provide conduction cooling instead of immersion cooling. Conduction cooling is generally free from logistic and major reliability problems. In an MRI system, the superconducting magnet is immersed into a liquid helium bath at a temperature of 4.2 K. The G-M cryocooler cools the radiation shield used to reduce the helium boil-off to a very low level, which can be easily maintained with infrequent helium delivery as seldom as once a year. Cryopumps generally use G-M refrigerators to cool the cryopumping surfaces to a cryogenic temperature as low as 20 K.

Accelerated research and development activities on conductively cooled magnets for MRI system applications and high-temperature superconducting materials are needed to improve the performance of G-M cooler-based MRI systems in terms of reliability and resolution. An MRI conductively cooled magnet employs a two-stage G-M refrigerator to cool both the magnet and the radiation shield, thereby requiring only a few watts of refrigerants. G-M/J-T crycoolers, when integrated with SQUID sensors, have potential applications in antisubmarine warfare (ASW) missions to detect submarines or underwater targets of military importance.

From 1960 to 1980, G-M/J-T cryocoolers were widely used in low-noise microwave amplifiers deployed in complex military equipment because of their high reliability and minimum refrigerant requirement during extended operations under severe environments. However, Stirling coolers were used mostly in military IR systems where the maintenance interval requirement exceeded 3000 hours. Both the MRI and cryopumping require two-stage (Fig. 4.11) cryocooling, one over the 80 to 20 K temperature range with a 40- to 50-W cooling power requirement and the other or second stage at 4.2 K with only 4 W of cooling power. G-M/J-T coolers need only yearly maintenance service, because of low helium loss

4. Critical Design Aspects with Low Cooling Capacities

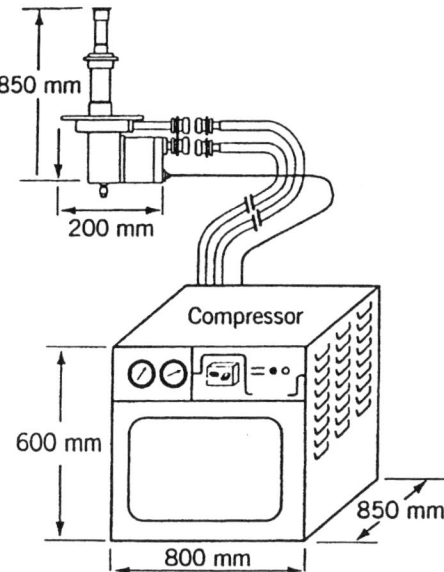

Fig. 4.11 Gifford-McMahon cycle, two-stage refrigerator for MRI applications. *Source: Superconductor Industry*, pp. 15–24, Fall 1993.

and minimum downtime. Since the helium loss and downtime have critical impact on the cooler reliability, they must be given serious consideration during the design phase of the cryocooler if maintenance-free cooler operation is the principal requirement.

4.7.7 G-M-CYCLE CRYOCOOLERS

G-M cryocoolers are the least expensive compared to G-M/J-T coolers. G-M-cycle cryocoolers are widely used in medical, scientific, and low-temperature physics experiments because of reasonably good performance and minimum cost. These coolers are also used in low-noise RF and MM-wave amplifiers and IR sensors used by the military because of the high maintenance interval which exceeds 3000 hours. There would be tremendous demand for G-M coolers in commercial, industrial, space, and military applications because of uninterrupted, maintenance-free operation over extended periods. Sales market projection data indicate that more than 2000 such units per year are manufactured and distributed worldwide for MRI applications.

4.7.8 COLLINS-CYCLE REFRIGERATOR SYSTEMS

The scientific and technical literature on cryocoolers indicates that Collins-cycle refrigerators offer the lowest operating temperature (4.2 K) with minimum cost and complexity. Such coolers are not suited for operations with low cooling requirements. The closed-cycle Collins refrigerator illustrated in Figure 4.12 is widely used by fusion magnet test facilities, high-energy accelerators, magnetically levitated trains, cryopumping systems, magnetohydrodynamic systems, and magnetic resonance imaging (MRI) systems [8].

4.7.9 HIGH-TEMPERATURE REFRIGERATOR SYSTEMS

High-temperature (77 K or higher) superconducting systems are designed to operate at higher cryogenic temperatures (in excess of 50 K) to meet specific performance requirements for various system applications. An acoustic sonar transmitter using high-temperature superconducting technology was developed under a Small Business Innovative Research (SBIR) contract [9]. The superconducting cooling system was designed to cool the sonar transmitter illustrated in Figure 4.13 for underwater target detection. This particular superconducting sonar transmitter configuration represents a classic example of integrating three distinct technologies: high-temperature superconductor (HTSC) technology, magnetostrictive technology, and cryocooler technology. This represents the first successful application of superconducting technology to an acoustic transducer without using liquid cryogen. The overall system architecture consists of magnetostrictive elements, a pair of magnetic coils wound from HTSC wire, and a Stirling-cycle refrigerator to maintain an operating temperature of around 50 K. The cooling level was maintained under both AC and DC currents flowing into HTSC coils without a liquid cryogen bath, thereby yielding maximum economy and reliability. It is important to mention that the cooling of the acoustic transducer must be efficient and reliable because it requires very high input power at low frequencies for accurate mapping and detection of underwater targets over distances exceeding 500 meters. Potential applications of cryogenically cooled sonar transducers include seismic tomography, global warming

4. Critical Design Aspects with Low Cooling Capacities

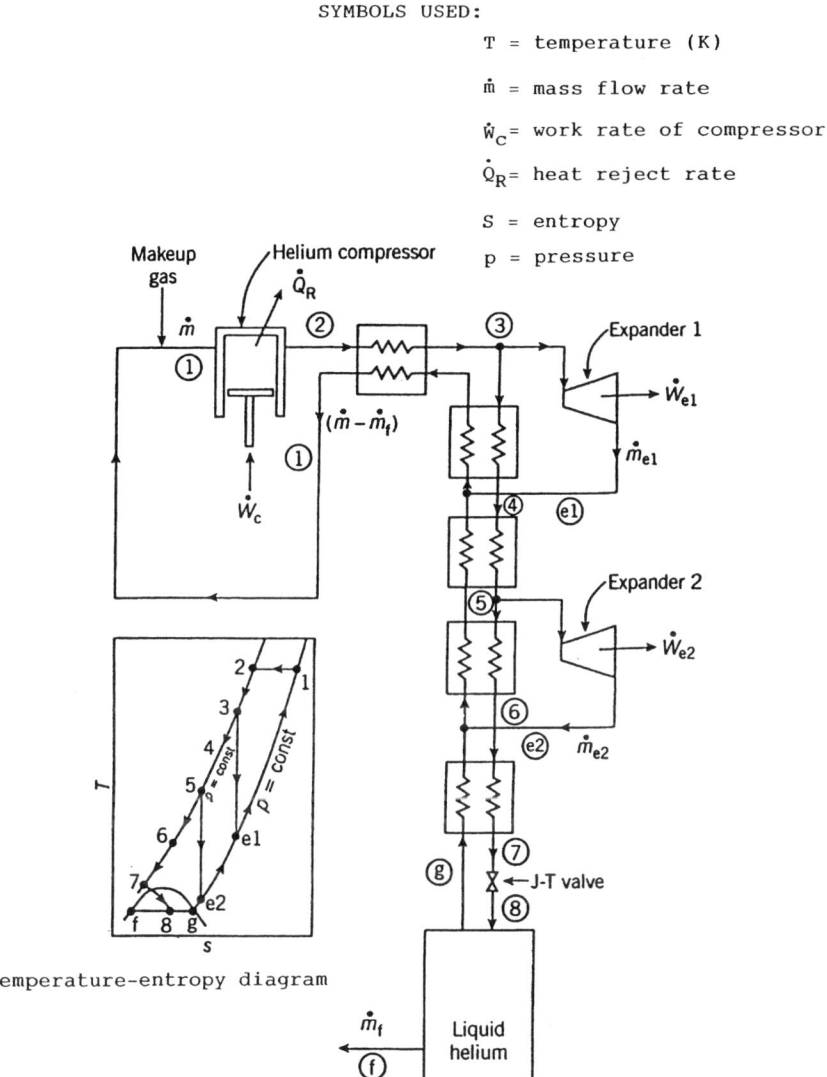

Fig. 4.12 Block diagram showing the critical elements of a closed-cycle Collins refrigerator system along with the T-S diagram [8].

measurements, ocean floor mapping, and ocean current monitoring [9]. The reliability of the cooling system used in this sonar transducer is inherently very high because of the few moving parts involved, low to moderate cooling requirements, and higher cryogenic temperature operation.

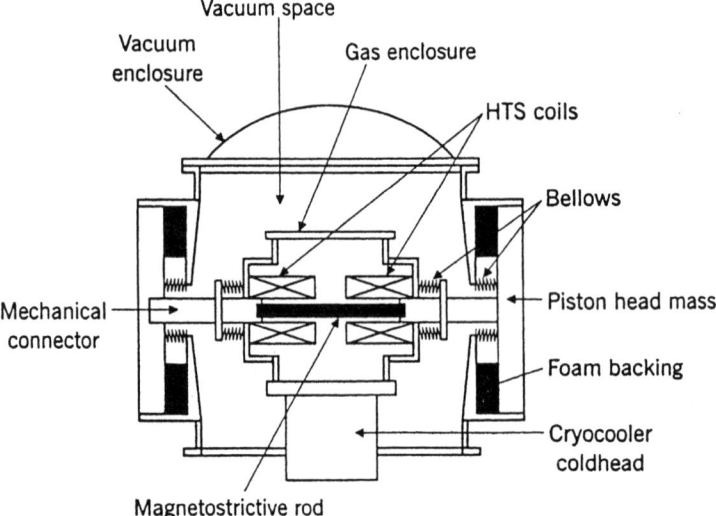

Fig. 4.13 Cross-sectional view of a high-temperature cooling system used for a sonar transmitter application [9].

4.7.9.1 Cooling Power Requirements at Higher Superconducting Temperatures

Cooling requirements at high superconducting temperatures between 50 and 75 K are not critical and can be easily satisfied with moderate input power. It is important to mention that the critical current in an HTSC wire increases linearly with the decrease in operating temperature, whereas the cooling power required by the cryogenic refrigerator follows an exponential relationship with the decreasing temperature. When the operating temperature is decreased from 77 to 50 K, the critical current is doubled, while the cooling power requirement increases by less than 15%. Cooling for a high-power system can be provided by a Stirling-cycle cryocooler designed for cryogenic electronics application with a typical capacity of 250 W at 77 K operation. This type of cryocooler has demonstrated a cooldown time less than 4 hours and reliability of more than 50,000 hours under severe mechanical and thermal environments. This particular cooler operating around 50 K is best suited for superconducting motors and generators, cryogenically cooled ship-propulsion systems, magnetic storage devices, and low-cost, portable MRI systems. Managing liquid helium cooling for the ship propulsion system shown in Figure 4.14 is not only

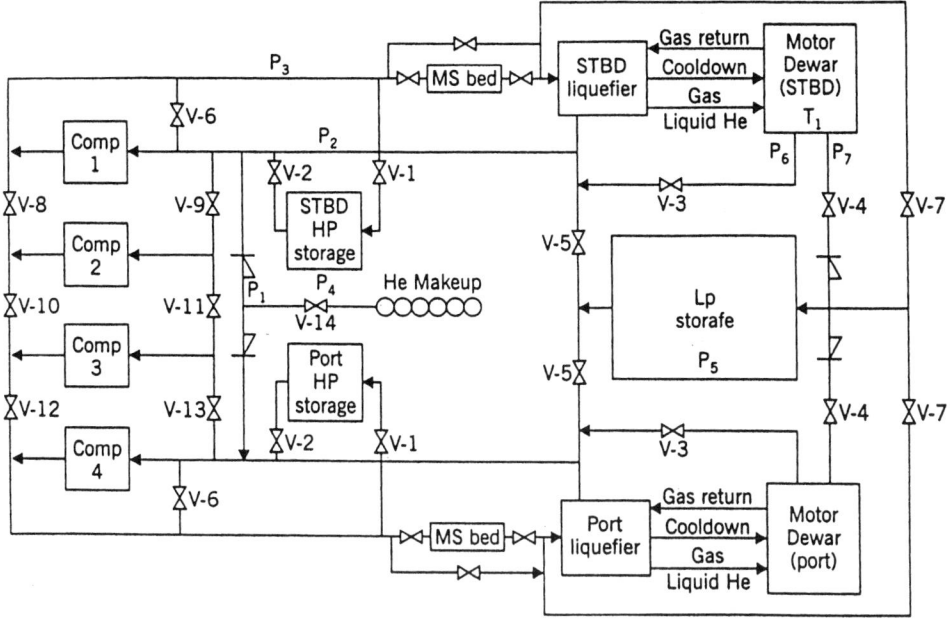

Fig. 4.14 Liquid helium supply and management system for ship propulsion application [8].

difficult but also complex. Critical components and operating parameters for liquid helium cooling managing systems are shown in Figure 4.14.

4.8 Performance Capabilities of Microcoolers

Microcoolers are best suited for providing cryogenic cooling for focal planar array (FPA) detectors, optical detectors, optical amplifiers, space sensors, military IR systems, and missile receivers. The weight, power consumption, cost, and size of microcoolers depend on the cryogenic operating temperature. Furthermore, regular maintenance and servicing of coolers for space sensors or tactical missiles are not possible due to logistic problems and thus a microcooler with effective integration of cryogenic electronics will be best suited for such applications. Studies performed by the author indicate that microcoolers intended for missile and space applications must meet high mechanical integrity, maintenance-free operation, ultra-high reliability, minimum power consumption, and virtually zero loss of cryogen. The studies further indicate that a conductive cooling scheme will be most attractive because of higher cooling capability, improved reliability, minimum weight, and higher thermodynamic efficiency.

4.8.1 POTENTIAL COOLING SCHEMES FOR MICROCOOLERS

The selection of a cooling scheme for a microcooler depends on the cooling capacity requirement and the type of cryogen to be used. Important characteristics of common cryogens are summarized in Table 4.2. The choice between immersion cooling and conduction cooling is a fundamental design issue for a microcooler or cryocooler and, therefore, the overall performance requirements and cost parameters must be taken into account before selecting any cooling scheme. In the case of immersion cooling involving liquid helium, one can deploy a Collins liquefier for a closed-cycle operation. An open-cycle operation with a liquid helium supply provision cannot be justified for space applications because of reliability and logistic problems. G-M/J-T microcoolers must be preferred in cryogenic systems with thermal losses not exceeding 5 W. Note that under transient state operation during the cooldown phase, the helium boil-off is allowed to escape from the cryostat and the cryostat is provided with fresh supply of liquid helium, which is again not feasible in space applications. For ground or large airborne system applications, frequent maintenance is desirable irrespective of the cooling schemes to keep the system operating with high efficiency and reliability over extended periods.

Thermodynamic studies performed by the author on various cooling schemes reveal that a conduction-cooling scheme eliminates the need for frequent helium refills, thereby yielding a most economical operation. Most commercial microcoolers with small cooling capacities not exceed-

Table 4.2 Typical Characteristics of Potential Cryogens for Microcoolers.

Cryogen	Boiling temperature (K)	Relative vaporization	Specific power (W/W)
Helium	4.2	1*	1000
Nitrogen	77.0	64	30
Neon	27.2	41	140

*Indicates 28 mw of power to boil 1 liter of liquid helium per day.

ing 5 W offer a single point of cooling. The use of such microcoolers or cryocoolers requires intimate familiarity with thermal load requirements, maximum allowable operating temperatures, and optimum thermal design to minimize the temperature differential between the superconductor and the microcoolers.

It is of critical importance that the cryocoolers or microcoolers must be designed to handle heat loads with adequate safety margins, particularly during the most critical operation. The transient-state operation presents greater thermal loads, which can limit the duty cycle of the cryogenic system. The environmental factors, such as ambient temperature, shock, and vibration, must be taken into account during the selection of a particular microcooler. A sophisticated high-capacity cooling system using liquid helium for a ship-propulsion application is shown in Figure 4.14. This cooling system offers high reliability over extended-period operations and provides safety interlocks to protect the system against high temperatures, contamination, insufficient cooling water, and low supply voltage. The optimum distribution of refrigeration capacity versus temperature is of critical importance in all microcooler designs, regardless of cooling capacity. As stated previously, a small liquid-helium cryostat or microcooler can support a total heat load of 29 mw at 4.2 K cryogenic temperature and still have the capability to absorb an additional load of 7.6 mw/K due to the heat capacity of the vapor. It is equally important to point out that thermal losses must be kept to a minimum if maximum efficiency and minimum operating costs are the principal requirements. Typical thermal losses in small cryocoolers or microcoolers are summarized in Table 4.3. In the case of multistage cryocoolers, heat loads and thermal losses must be carefully distributed to achieve optimum cooler operation.

Table 4.3 Typical Thermal Losses for a Small Conductively Cooled Refrigeration System.

Thermal loss type	*First-stage loss*	*Second-stage loss*
Operating temperature (K)	40	10
Conduction/radiation (W)	17.2	0.4
Power leads (W)	14.0	2.1
Thermal margin allowed (W)	7.8	0.6
Total thermal losses (W)	39	3.1

4.9 Performance Capabilities and Limitations of Microcoolers

Miniaturized versions of cryocoolers are known as microcoolers, which have potential applications to space sensors and military airborne systems such as IR line scanners, IR search and track sensors, high-resolution thermal imaging systems, and forward-looking IR (FLIR) sensors. Microcooler design based on the Stirling-cycle operating principle can achieve a cooling temperature of 77 K within 3 minutes. The microcooler configuration normally incorporates a generator to produce a compression-expansion refrigerator cycle with no valves. It is important to mention that the regenerator has a large heat capacity and acts like an efficient heat exchanger. A microcooler designed for NASA thermal imaging applications consumes electrical power less than 3 W, weighs less than 15 oz, has a life expectancy of 5 years, and has demonstrated continuous operation exceeding 8000 hours with no degradation in performance. High reliability of a microcooler requires improved materials with self-lubrication capability, elimination of gaseous contamination, clearance seals with very low friction, and a linear drive mechanism. Design analysis performed by the author indicates that a microcooler can provide a cooling capacity of 150 mw at 77 K cryogenic temperature with input power less than 3.5 W, heat load of 150 mw, and cooldown time ranging from 1.5 minute to 120 K to 4 minutes to 77 K from the room temperature (300 K). Typical thermal losses and operating temperatures for the first and second stage are summarized in Table 4.3 for a small conductively cooled system with higher cooling capacities.

High-performance microcooler designs using rare-earth regenerator materials have demonstrated significant improvement in thermal performance and reliability. A 4.2-K, two-stage GE microcooler using erbium-nickel rare-earth materials as regenerators has attained a cryogenic temperature of 4.2 K in a remarkably short time because the erbium nickel material has improved heat capacity below a temperature of 10 K. This particular microcooler is best suited for IR missile seeker receivers and battlefield tank sensors, where time is of critical importance. The heat load capacity of this microcooler is about 100 mw at 4.2 K operation, but a heat load capacity exceeding 1 W has been demonstrated at higher cryogenic temperatures under controlled environments and employing unique combinations of rare-earth materials.

4.10 Specific Weight and Power Estimates for Cryocoolers

Specific weight (lb/W) is defined as the cryocooler weight per watt of input power, whereas the specific power (W/W) is defined as the input power in watts per watt of heat load or cooling power. In other words, specific power is the ratio of input power to cooling power required. Both the specific weight and specific power estimates vary from cooler to cooler and can reveal the complexity of the cooling system instantly. Typical values of specific weight and power for heavy-duty cryogenic systems such as MRI or sonar transmitters using Stirling-cycle cryocoolers [10] are illustrated in Figure 4.15. Performance characteristics such as operating temperature, cooling power, input power, and specific power and weight of J-T-closed-cycle refrigerator systems are shown in Table 4.4 along with appropriate comments under the remark column.

Typical characteristics of Stirling-cycle refrigerator systems such as cooling power, input power, specific power, weight, and specific weight are summarized in Table 4.5.

Table 4.4 Performance Characteristics of Joule-Thomson Closed-Cycle Refrigerator Systems [10].

System	Operating temperature (K)	Cooling power (W)	Input power (W)	Weight (lb)	Remarks
1	77	2.25	450	25	Ram air cooling used
2	77	5	650	22	Includes fan cooling power
3	77	3	450	19	Ram air cooling used
4	77	2	460	23	Includes fan power
5	77	2	650	23	Includes fan power
6	77	2.5	530	20	Includes fan power

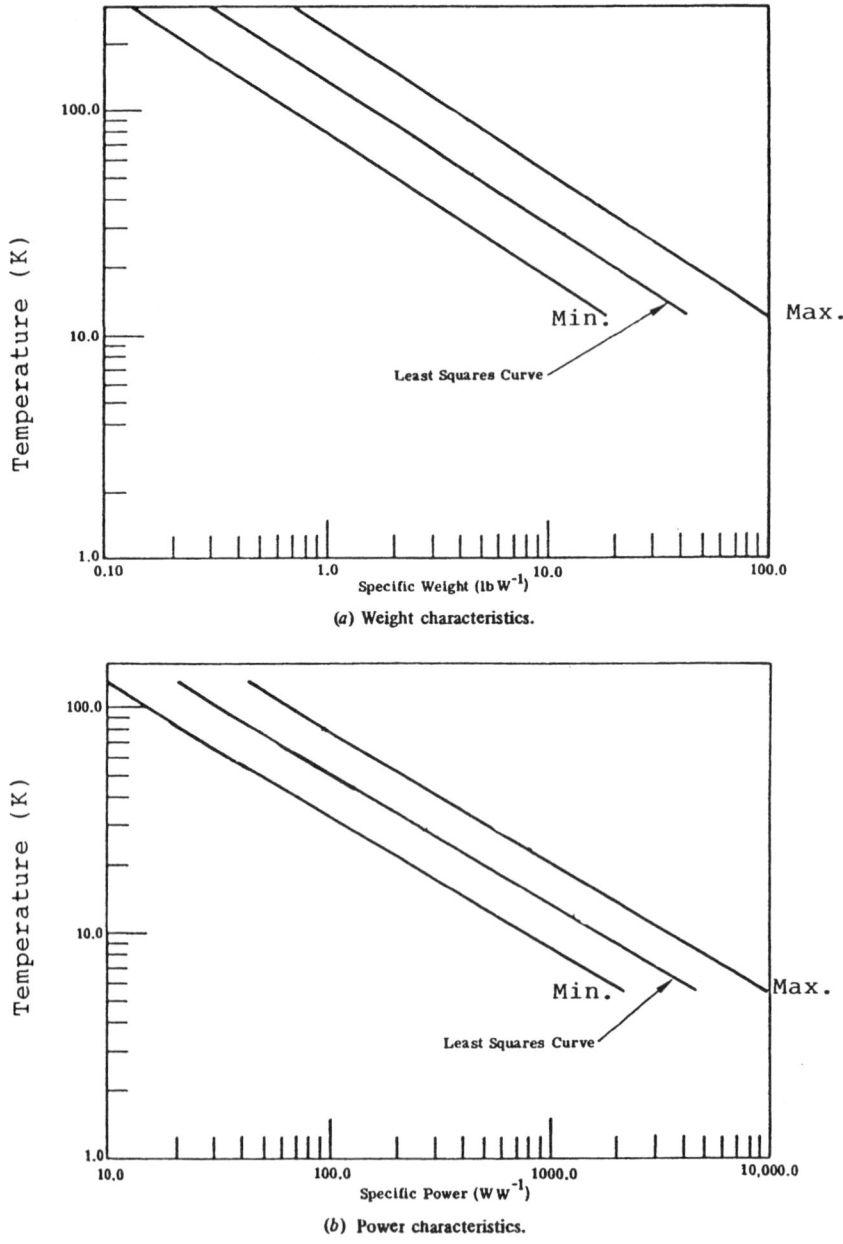

Fig. 4.15 Weight and power characteristics of a Stirling-cycle refrigerator system.

Table 4.5 Typical Performance Characteristics of Stirling-Cycle Refrigerators [10].

System	Temperature (K)	Cooling power (W)	Power Input (W)	Power Sp. power (W/W)	Weight Total (lb)	Weight Sp. wt (lb/W)
A	25	2	480	240	15.5	7.75
B	60	1.5	120	80	5.5	3.7
C	30	1	620	620	16	16
D	20	10	1750	175	112	11.2
E	140	1.5	30	20	15	10
F	25	1	850	850	29	29
G	77	3.5	180	51	6.5	1.85

Performance characteristics summarized in Tables 4.4 and 4.5 provide instant information on the critical performance parameters on two of the most prominent refrigerator systems widely used in commercial, industrial, and military applications.

4.11 Thermodynamic Aspects and Efficiency of Cryocoolers

Thermodynamic aspects must be given serious consideration because they have a significant impact on the thermodynamic efficiency of a refrigerator system, regardless of its cooling capacity. For specific details on the thermal aspects, readers are referred to textbooks on refrigeration system technology with an emphasis on the thermal design parameters such as operating temperature, cooling capacity, and input power requirement. Thermodynamic efficiency of a cryocooler with 1-W cooling capacity and operating at a cryogenic temperature of 4.2 K can be determined by looking at the power consumption with respect to the power consumption in the ideal Carnot cycle [4]. Theoretically, the input power required per watt of refrigeration or cooling capacity at the 4.2 K cryogenic temperature is about 70 W. A performance survey of the commercially available G-M/J-T cryocoolers indicates that an input power of 4500 W is required to produce a cooling power of 1 W at the cryogenic temperature of 4.2 K and the ideal efficiency of such cryocoolers is about 1.6%. However, a G-M-cycle cryocooler using 525 grams of rare-earth material would require an input power close to 7000 W to produce 1.05 W of cooling power at a temperature of 4.2 K with a Carnot cycle efficiency of 1.1%.

4.11.1 THERMAL ANALYSIS OF REFRIGERATION SYSTEM

Cooling systems can be classified into two major categories: mechanical-powered refrigerator cooling systems (Fig. 4.16[A]) and heat-powered refrigerator cooling systems (Fig. 4.16[B]). In the case of a mechanical-powered refrigerator, the heat is absorbed at cooler temperature (T_c) but rejected at higher ambient temperature (T_a) as illustrated in Figures 4.16(A) and 4.16(B), where temperature T_a is greater than T_c. The expressions for the work required (W_o) for a mechanically driven refrigerator and the thermal energy required for a heat-driven refrigerator (q_h) can be written as follows:

$$W_o > q_c[(T_a - T_c)/T_c] \quad (4.1)$$

$$Q_h > q_c[T_h/T_c][(T_a - T_c)/(T_h - T_a)] \quad (4.2)$$

where q_c is the thermal energy cold source (Joules), q_h is the thermal energy for the heat-driven refrigerator (Joules), T_h is the absolute temperature of the heat source (K), and W_o is the mechanical work required (Joules or watts-sec).

It is important to mention that the input power is at minimum when all the processes of the thermodynamic cycle are reversible — that is, entropy S is equal to zero (S = 0). Furthermore, reversible cycles can be achieved in theory, but practical refrigeration cycles are irreversible. The figure-of-merit (FOM) or the coefficient of performance (COP) is the most critical performance parameter of a cryogenic refrigerator [10]. The COP is defined by the ratio of refrigeration or cooling power produced to input power supplied. In other words, the specific power of a refrigeration system is an inverse of COP. Based on this statement, one can now write

$$\text{Specific power } (P_{\text{specific}}) = [1/\text{COP}] \quad (4.3)$$

Understanding various heat-engine cycles is of paramount importance if a meaningful performance comparison is desired between normal and reverse operations. Note that the Carnot cycle is used as a standard of comparison for heat-engine cycles because its efficiency is the maximum for a given temperature limit. However, the reversed-Carnot cycle has been widely accepted as a standard of comparison for refrigeration operating cycles, because its COP is maximum for a given temperature limit. The temperature-entropy (T-S) diagram for the reversed-Carnot cycle

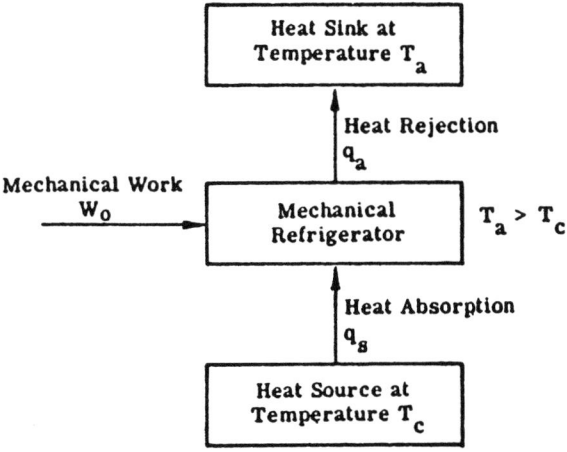

(A) Mechanical Refrigerator

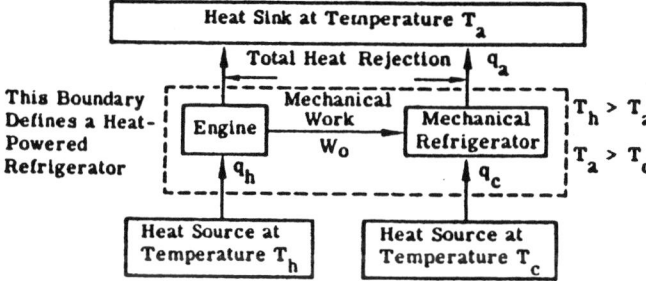

(B) Heat-driven Refrigerator

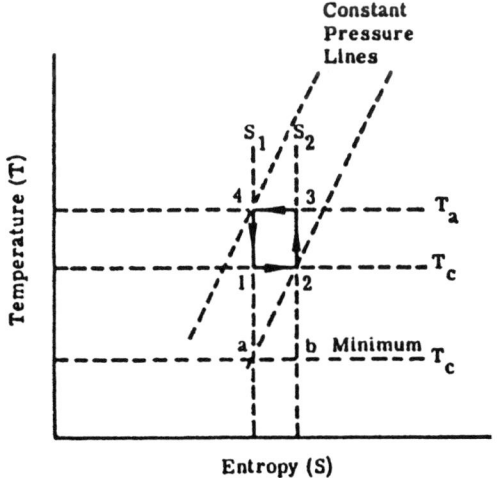

(C) Temperature-entropy diagram

Fig. 4.16 Refrigerator types, (A) mechanical, (B) heat-driven, and (C) temperature-entropy diagram for reversed-Carnot cycle.

shown in Figure 4.16(C) indicates the heat absorbed at temperature T_c during the process 1–2 represented by the area 1-2-b-a-1, while the heat rejected at the ambient temperature T_a during the process 3–4 is represented by area 3-4-a-b-3. The net work input is represented by the difference between these two areas and is shown by the area 1-2-3-4-1 in Figure 4.16(C).

The compression, cooling, expansion, and heating processes are accomplished isothermally (i.e., under constant temperature conditions). Note that the heat transfer processes during these phases are affected even over slight temperature differences, resulting in no overall increase in entropy. The fluid is cooled and heated between these temperatures by isentropic or constant entropy expansion and compression, respectively.

4.12 Weight Requirements for Cyrogens Used by Cryocoolers

The expression for the ideal weight of a cryogen regardless of forms (solid or liquid) is dependent on the cryogenic properties of the cryogen used, the mission time requirement, and the environmental conditions including the dynamic effects, if any. The cryogen weight estimation expression generally assumes the following:

- The surface area for the heat transfer from a spherical cryogen-Dewar is simply assumed to be πD^2, where D is the diameter.
- The Dewar is insulated with a multiplayer insulation (MLI) that has an effective emissivity of ε_{eff}.
- The heat leakage into the cryogen is through the MLI.
- The heat absorbed by the cryogen can be represented by the product of heat leakage and total elapsed time, t_e.

The weight of the cryogen under these assumed conditions can be expressed as

$$W_c = \left[113/\delta^2\right]\left[(\sigma\varepsilon_{\text{eff}})(t_e)(T_s^4 - T_c^4)/h_c\right]^3 \qquad (4.4)$$

where W_c is the weight of cryogen (gram); δ is the density of the cryogen (gram/cm³); σ is Stefan-Boltzmann constant equal to 5.6703×10^{-12} (W/cm²/K⁴); ε_{eff} is the effective emissivity (dimensionless unit); T_s and T_c

are the absolute temperature of the outer shell of the vessel containing the cryogen and the absolute temperature of the cryogen in Kelvin (K), respectively; t_e is the total elapsed time (sec); and h_c is the latent heat (or heat of vaporization for a liquid) of the cryogen (W-sec/gram).

The constant equal to 113 is for the spherical geometry assumed. Note that this constant is strictly dependent on the storage tank geometry. Relevant data on various cryogens, dry weights of cryogenic-liquid Dewars, representative data on solid-cryogens, vessel weights, the heat transfer for the MLI systems, and effective emissivity may be obtained from Chapter 15 in reference [10]. The author is suggesting the following approximate values for various parameters:

1. Effective emissivity:
   $\varepsilon_{\text{eff}} = 0.009$ for $N = 2$, 0.004 for $N = 5$, and 0.002 for $N = 10$, where N is the umber of isolated insulation layers using double-aluminized Mylar material with different emissivity values
2. Dry weight of cryogenic-liquid dewars:
   167 lb for Inconel material, 126 lb for aluminum-alloy 2219, and 115 lb for titanium
3. Solid-cryogenic vessel weight:
   27 lb for nitrogen at 58 K cryogenic temperature

4.13 Charactcrisitcs and Storage Requirements for Potential Cryogens

Cryogens are available in three distinct forms: gas, liquid, and solid. Cryogenic liquids or fluids when stored in equilibrium with their vapors (known as subcritical storage) can provide a constant-temperature control system. Cryogenic liquids are available with temperature capabilities ranging from 4.2 to 240 K. The primary limitation of this approach is the complex tank design requirement needed to minimize the boiloff, the direct relation of weight and volume requirements to the elapsed time, and phase separation in the weightless or space environment. However, cryogenic liquids can be stored at pressures well above their critical pressures (known as supercritical storage) as homogeneous fluids, thereby eliminating the phase-separation problems frequently encountered during the weightless or space environment. The high pressures of supercritical

Table 4.6 Performance Capabilities and Limitations of Various Cryocoolers.

Cryocooler category	Cooler type	Temperature range (K)	Cooling capacity (W)	Coefficient of performance (COP)	Remarks
(1) Open-cycle	Subcritical	4.2–77	Unlimited	NA	Reliable operation complex Dewar
	Supercritical	5.2–126	Unlimited	NA	Complex design
	J-T expansion	4.2–87.4	20 W (max)	NA	High gas purity
(2) Mechanical refrigerators	VM	10–77	0.020–15	0.02 (max)	Limited data available
	Stirling	10–77	0.020–20	0.05 (max)	Compact and minimum consumption
	G-M	10–27	0.015–20	0.01 (max)	Best suited for airborne sensors
	Closed cycle Joule-Thomson	50–77	0.5–5	0.01 (max)	Limited life, high input power

Note: NA stands for not available.

storage require heavier containers with high mechanical integrity needed under subcritical storage.

4.14 Classifications of Cryocoolers and Their Performance Capabilities and Limitations [10]

The cryocoolers can be classified into two major categories as follows:

1. *Open-cycle cryocoolers.* These are expandable cooling systems that use stored cryogens in either the subcritical or supercritical liquid state; solid cryogens; or stored, high-pressure gas with a Joule-Thomson (J-T) expansion valve [10].
2. *Closed-cycle, mechanical cryocoolers.* These cryocoolers provide cryogenic cooling at very low cryogenic temperatures and reject the heat at high temperatures. Mechanical refrigerator systems include Vuilleumier (VM) coolers, Stirling coolers, Gifford-McMahon (G-M) coolers, Brayton-cycle coolers, and closed-cycle Joule-Thomson (J-T) coolers. Performance capabilities and limitations of these two types are summarized in Table 4.6.

4.15 Summary

This chapter discussed the design aspects of cryocoolers used in commercial, scientific, space, and military applications, with an emphasis on reliability and thermodynamic performance. Closed-cycle cryocoolers were described in greater detail because of their proven performance including their cost-effective operation, stealthy design, and reliability. The performance capabilities of Stirling cryocoolers, widely used in space and military applications, were summarized with emphasis on reliability during extended operations. Collins and G-M refrigerator systems have demonstrated greater suitability for magnetically levitated trains, magnetic resonance imaging (MRI) systems, and cryovacuum pumping applications. The maintenance aspects of cryocoolers were briefly discussed with particular emphasis on reliability, loss of cryogen, and freedom of refill. Cooling capacity curves as a function of heat load and operating temperatures were examined to give instant information on cooling capacity. Boreas-cycle cryocoolers, which use both the high-pressure ratio and counterflow heat exchanger, were described with specific details on performance level and reliability. The advantages of high-pressure ratio expansion include improved efficiency, minimum power consumption, low operating temperature, high reliability, reduced helium mass flow, lower cold-head speed, and longer compressor life. Temperature capability, stability, and the mass flow rate for selected cryocoolers were identified under specific operating environments. Microcooler and cryocooler designs incorporating advanced material technology and unique heat exchanger configurations were discussed. Performance capabilities and limitations of widely used cryocoolers such as Stirling-cycle, G-M-cycle, J-T-cycle, and Boreas-cycle coolers were briefly described with an emphasis on heat lead, reliability, and input power requirement. Potential cooling schemes — namely, conduction, convection, and radiation — were discussed with an emphasis on temperature differential and thermal efficiency. Specific weight, specific power, and coefficient of performance (COP) for various open-cycle and closed-cycle refrigerators were examined for the benefit of readers interested in the rapid selection of a cryocooler for specific application. Mathematical expressions for work done by compressor, thermal energy for the heat source, and weight of cryogen were derived in terms of several variables including cryogenic temperature, heat energy, density, effective emissivity, and latent heat of cryogen. Performance capabilities, limitations, and coefficients of performance for widely used cryocoolers were summarized in Table 4.6.

References

1. A. R. Jha. *Infrared technology: Applications to Electro-optics, Photonic Devices, and Sensors*, New York: John Wiley & Sons, August 2000, pp. 427–428.
2. A. R. Jha. *Superconductor technology: Applications to Microwave, Electro-optics, Electrical Machines, and Propulsion Systems*. New York: John Wiley & Sons, 1998, p. 290.
3. Staff Reporter. "Authropological information." *Photonics Spectra*, September 1998, pp. 91–98.
4. Jha, *Superconductor technology*, New York: p. 288.
5. Jha, *Infrared technology*, p. 430.
6. A. Bitterman et al. "Pulse tubes foothold in cryocooler markets." *Superconductor and Cryoelectronics*, September 1998, pp. 12–17.
7. Jha, *Superconductor technology*, pp. 280–284.

8. R. A. Auckermann. " Closed-cycle refrigeration for superconducting applications." *Superconductor Industry*, Fall 1993, pp. 15–24.
9. C. L. Joshi et al. "Putting a chill into the HTSC applications." *Superconductor Industry*, Fall 1993, pp. 26–29.
10. W. L. Wolfe and G. J. Zissis. *The Infrared Handbook.* Environmental Research Institute of Michigan, 1978, Chapter 15, pp. 27, 31, 39, and 41.

Chapter 5 | Performance Requirements for Moderate- and High-Capacity Refrigeration Systems

5.0 Introduction

This chapter focuses on the critical design aspects and performance capabilities of moderate- and high-capacity refrigeration systems. High-capacity refrigeration systems include high-pressure gas refrigeration systems and turbine-based cooling systems, known as turbo-pumps. Moderate-capacity refrigerator systems are used in scientific research and medical diagnostic applications. Important design issues and critical performance parameters of these cooling systems are discussed in this chapter with particular emphasis on reliability, cooling capacity, cooling efficiency, mass flow rate, and overall performance as a function of various operating parameters. Design aspects of heavy-duty mechanical refrigeration systems are briefly summarized with emphasis on cost and reliability.

5.1 Description of High-Capacity Refrigeration Systems

High-capacity refrigeration systems are best suited for commercial and industrial applications. Such systems generally operate at high pressures to achieve high cooling capacity. A high-capacity refrigeration system (HCRS) uses high-pressure gas as a cooling agent. A classical example of a high-pressure refrigerator system based on the Joule-Thomson (J-T)-cycle principle is illustrated in Figure 5.1. This figure shows the temperature-entropy (T-S) diagram along with a system schematic diagram identifying the critical elements of the system. It is important to point out that this particular refrigerator cycle is identical to the reversed-Brayton cycle except for one fundamental difference [1], which states that the expansion procedure (4–5) is accomplished by the

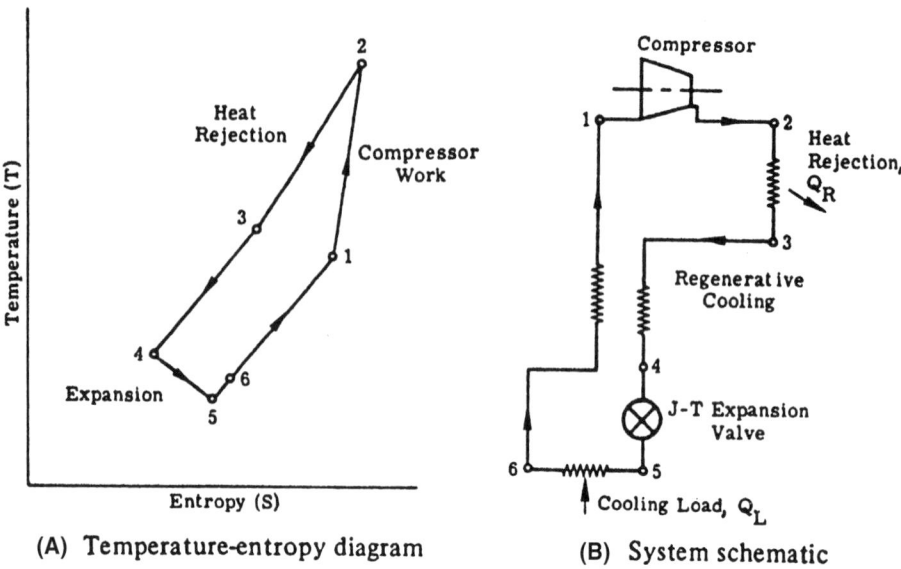

Fig. 5.1 System schematic and temperature-entropy (T-S) diagram of a J-R-cycle refrigeration system.

expansion through a throttling valve in this case rather than a turbine as illustrated in Figure 5.2 for a reversed-Brayton-cycle refrigeration system.

In the case of the J-T-cycle refrigeration system, operating point 5 represents a two-phase region. The heat of vaporization of the coolant is utilized to absorb the heat from the cooling loading during the short duration process (5–6). The working cooling agent is generally nitrogen gas, which is compressed to a high-pressure close to 250 psia in a multistage, oil-lubricated, reciprocal compressor unit shown in Figure 5.1. The heat generated by the compressor unit is removed by RAM air or by a fan attached to the compressor assembly. The compressed nitrogen gas passed through an absorption filter (not shown in Fig. 5.1) that eliminates the oil vapor and other contaminants has a tendency to solidify at cryogenic temperatures. The purified, high-pressure nitrogen gas enters a miniature regenerative heat exchanger known as a cryostat, where it is cooled by the returning low-pressure nitrogen gas. At the output of the heat exchanger, the high-pressure nitrogen gas is expanded through an expansion valve, leading to a significant temperature drop, sufficient to liquefy a portion of the nitrogen gas. The latent heat of the liquid nitrogen is used to provide

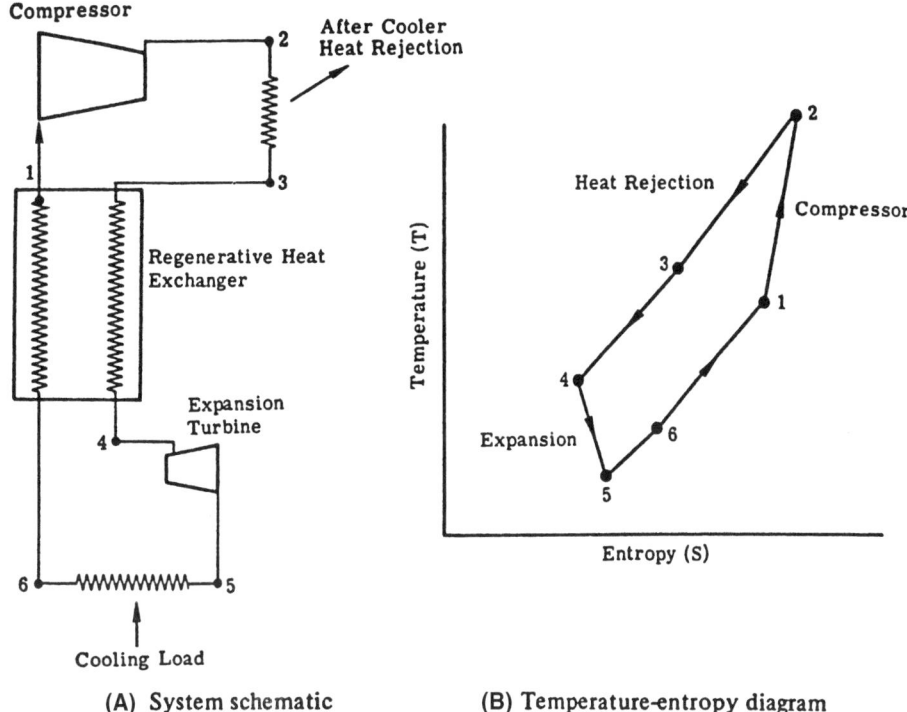

Fig. 5.2 System schematic showing the critical elements of the system and temperature-entropy (T-S) diagram for a reversed-Brayton-cycle refrigeration system.

spot cooling. The low-pressure gas is returned to the first stage of the compressor. A compact gas reservoir (Fig. 5.1) known as an accumulator is connected to the low-pressure return gas line to adjust the gas supply or volume to compensate for the increased density of the working fluid that is liquefied during the normal system operation.

Working fluids such as nitrogen and argon are widely used in J-T-cycle refrigeration systems involving high-pressure operations. The cooling capacity (watt per liter per minute of gas flow) for these two fluids as a function of operating pressure is illustrated in Figure 5.3. Preliminary calculations indicate that the maximum cooling capacity for nitrogen occurs at an operating pressure in the vicinity of 5000 psia, whereas for argon it occurs at a pressure of 8000 psia as shown in Figure 5.3. One can see that argon offers more than 65% cooling capacity over nitrogen, but at significantly higher operating pressures. Tradeoff studies must be performed

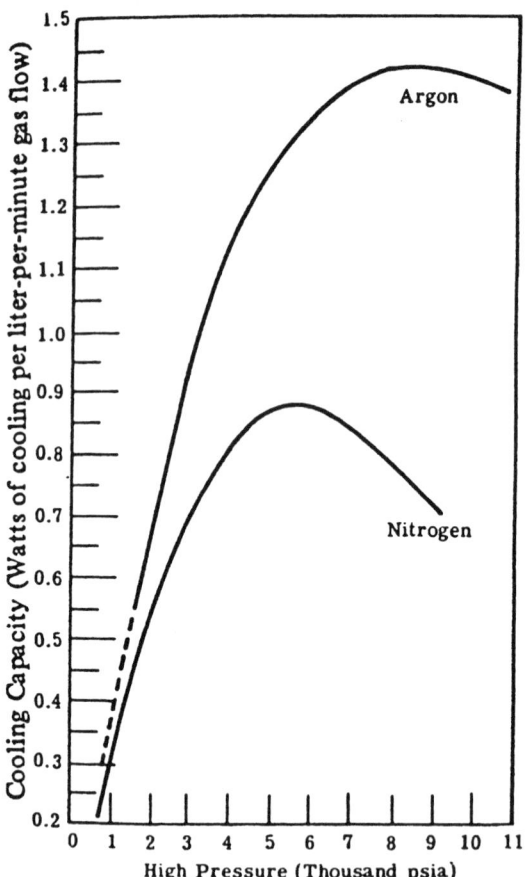

Fig. 5.3 Cooling capacity as a function of operating pressure for the J-T-cryogenic system deploying nitrogen and argon as working fluids.

to achieve cost-effective cooling operation, high cooling capacity, and improved reliability at higher operating pressures. High-capacity refrigeration systems are widely used in industrial, manufacturing, and space rocket launch applications.

5.1.1 CLAUDE-CYCLE REFRIGERATION SYSTEM

A Claude-cycle refrigeration system is similar to a Brayton-cycle refrigeration system except that an additional heat exchanger and a J-T valve

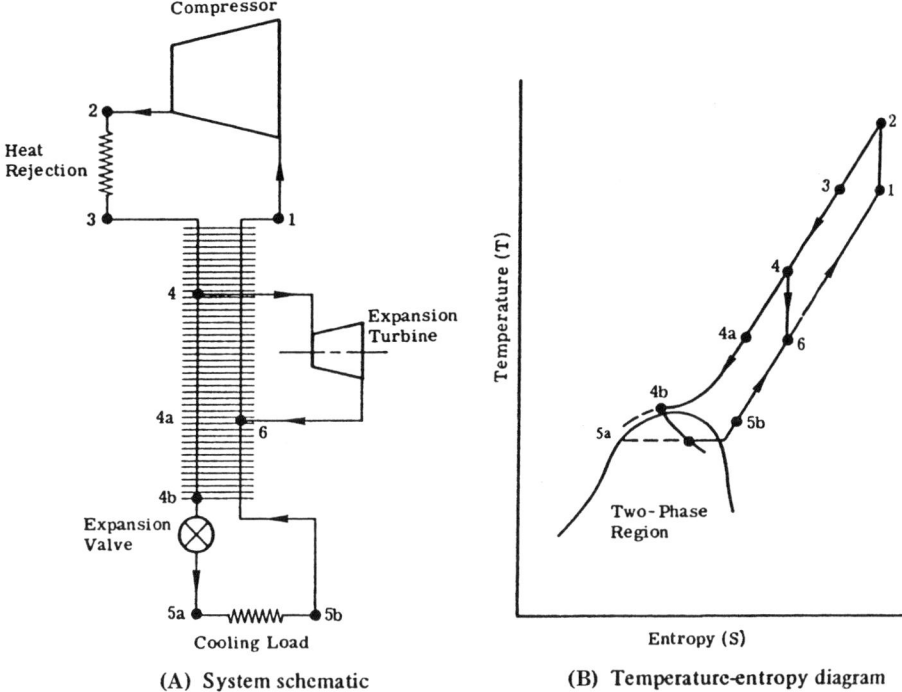

Fig. 5.4 System schematic and temperature-entropy (T-S) diagram of the Claude-cycle refrigeration system along with its critical system components such as the turbine, the compressor, and the expansion value.

are incorporated at the cold end to obtain an ultra low cryogenic temperature. In other words, the Claude-cycle is effectively a J-T-cycle refrigeration system in which the net heat sink temperature is reduced using a Brayton-cycle refrigerator. A system schematic, a temperature-entropy (T-S) diagram, and critical elements of a Claude-cycle refrigeration system are shown in Figure 5.4. The most critical elements of a Claude-cycle refrigeration system include a high-pressure, multistage compressor; a multistage expansion turbine; and an expansion valve. It is important to point out that a rapid temperature rise occurs in the temperature-entropy diagram as illustrated in Figure 5.4.

5.1.2 REVERSED-BRAYTON-CYCLE REFRIGERATION SYSTEM

High-pressure versions of reversed-Brayton-cycle refrigeration systems have been in service for several years, particularly in large ground-based installations. When using a turbine, this type of refrigeration system is referred to as a turbine-based refrigeration system. Critical system elements and a temperature-entropy (T-S) diagram of the reversed-Brayton-cycle system were shown in Figure 5.2. However, recent design efforts have been directed toward the development of gas-bearing supported turbo-machines and rotary-reciprocating mechanisms that are most suitable for the use in compact and efficient cryogenic refrigeration systems. A turbo-refrigerator system employing gas-bearing turbo-machinery composed of a gas turbine and a high-speed, multistage, high-pressure compressor unit offers high reliability and maintenance-free operation over extended periods leading to a most cost-effective design. Note that lubrication of the bearings with cycle working fluid (nitrogen or helium) excludes certain lubricants that cause contamination and fouling in the low-temperature region of the cooling cycle. The absence of contaminants eliminates the machine failure due to wear and tear. This clearly indicates that a cryogenic system based on a gas-bearing, turbo-machinery operating principle will provide long operating life unmatched by any other cryogenic system currently under operation.

5.2 Refrigeration Systems with Moderate Cooling Capacity

The following refrigeration or cooling systems are best suited for applications where moderate cooling is required:

- Gifford-McMahon (G-M)-cycle refrigeration system
- Joule-Thomson (J-T)-cycle refrigeration system
- Brayton-cycle refrigeration system

5.2.1 G-M-CYCLE REFRIGERATION SYSTEM

The G-M-cycle and the Solvag-cycle refrigeration systems basically operate on the same principles, but with different expander modifications to meet specific performance requirements. By separating the expander unit from the compressor, one can obtain a lightweight, compact cooling system that can be easily integrated with the cooling load. However, the compressor unit can be located separately at a convenient location. This way the compressor can be connected to the expander unit through long, flexible lines carrying both the high-pressure and low-

Table 5.1 Critical Design and Performance Parameters of Selected G-M-Cycle Refrigeration Systems.

System number	Temperature (K)	Cooling power (W)	Input power (W)	Input/ Refri. (W/W)	Weight (lb)	Weight (lb/W)
A	7.5–25	1 @ 9.5 K	3000	3000	200	200
B	10–28	2 @ 13 K	6100	3050	458	229
C	77	0.6 @ 77 K	135	225	6.3	10.5
D	23–89	10 @ 30 K	3000	300	200	20
E	30–77	1 @ 77 K	368	368	14.5	14.5
F	19–30	1 @ 26 K	800	800	25	25

pressure working fluids. Furthermore, all such cryogenic systems come with the hermetically sealed compressor units. The cryogenic cooling system configured this way offers low maintenance, high reliability, long operating life, and low operating costs.

There are several versions of G-M-cycle refrigerator systems on the market, and practically all use hermetically sealed compressor units to avoid the leakage of working fluid. However, the method of operating the expander unit and deployment of a high-pressure gas storage tank with high reliability and maximum safety will determine the procurement cost of this type of refrigeration system. Critical design parameters and performance capabilities of G-M-cycle refrigeration systems [1] are summarized in Table 5.1.

G-M-cycle refrigeration systems are referred to as mechanical refrigeration systems in some books. Note that the cooling capacity of a refrigeration system depends on the operating pressure and the cryogenic temperature desired. The cooling capacity is expressed in watts of cooling per liter per minute of gas flow. Gas flow is expressed in minutes @ one atmosphere pressure (760 mm of mercury or 14.7 psi) and at 0°C or at 273 K temperature. Cooling capacity for two distinct commonly used gases — namely, nitrogen and argon used in J-T coolers operating at 300 K temperature and as a function of gas pressure — is illustrated in Figure 5.1. As stated earlier, gas flow is a standard unit expressed in liters per minute at one atmospheric pressure (760 mm of mercury or 14.7 psi) and at a temperature of 273 K. An optimum safety factor must be used in designing the fluid storage tanks to ensure high mechanical integrity, particularly

Table 5.2 Typical Performance Parameters of High-Pressure Gas J-T Cooling Systems with Moderate Capacity [1].

System parameters	System 1	System 2	System 3	System 4
Cooling load (W)	7	6	10	0.20
Temperature (K)	80	22	80	87
Working fluid	N_2	LN_2	N_2	A
Gas consumption (L/min) At a specified pressure	22.6	0.008	14	1.2
Operating pressure (psia)	1160	1160	1030	600
Cooldown time (min)	5	10	0.50	0.33
Orifice type	Fixed	Fixed	Fixed	Variable
Overall dimensions (in) Length/diameter	0.5/3.5	17/6 (pre-cooler)	1.7/0.3	2.75/0.25
Application	Research	Research	Missile	Missile

when the cryogenic systems are operating at pressures 5000 psia and above.

5.2.2 J-T-CYCLE REFRIGERATION SYSTEM

This particular system is widely used in scientific investigation, space research, and missile applications where cooldown time, gas consumption, and cooling capacity are the most demanding requirements. Typical performance parameters of high-pressure gas-based J-T cooling systems are summarized in Table 5.2. Typical cooldown time for such refrigeration systems varies from 30 seconds to 10 minutes depending on the working fluids deployed by the system. Estimated values of cooling capacity of the nitrogen- and argon-based refrigeration systems as a function of gas flow rate and operating pressure are shown in Figure 5.3.

5.2.3 BRAYTON-CYCLE REFRIGERATION SYSTEM

The term "Brayton-cycle" is generally used to indicate a class of thermodynamic cycles that is made up of variants of the reversed-Brayton, thermodynamic cycle including the Claude-cycle system. A reversed-Brayton-cycle refrigerator system consists of a compressor unit, a heat exchanger assembly, an after-cooler unit, and an expansion turbine subsystem. It is important to mention that the compressor unit compresses the ambient-temperature (300 K) gas, and the heat is rejected to the ambient environmental temperature. The high-pressure gas is then passed through a regenerative heat exchanger and finally is expanded across the gas turbine where the heat is extracted. The gas is directed through the regeneration load or cooling load.

5.3 Turbo-Machinery Refrigeration Systems

Turbo-machinery refrigeration systems are best suited for commercial and industrial applications, where high cooling capacity and cost-effective system performance are the principal requirements. The cooling system's performance depends on the cooling capacity, the working fluid flow rate, and gas flow rate as the cooling capacity and flow rates are increased, the turbo-machinery design becomes most attractive in terms of size, weight, and performance. Typical power consumption and system weight estimates as a function of cooling capacity for a turbo-machinery refrigeration system are illustrated in Figure 5.5 for two distinct cryogenic operating temperatures. It is important to mention that both the input power (or power consumption) and system weight increase with the increase in cooling capacity. Furthermore, significantly large amounts of input power are required at a cryogenic operation at 4.4 K compared to an operation at 20 K. In brief, cost, system complexity, power consumption, and weight all increase at lower cryogenic temperatures.

It is interesting to mention that both the Stirling and Brayton cycles can be operated in "backward" or reverse direction to act as cooling machines or heat pumps. The cycle diagrams will not be identical to the parent heat engine cycles when the effects of turbine and compressor efficiencies are taken into account. The irreversibilities cause an increase in entropy, which changes the shapes of various process lines in the cooling cycles from those found for the heat engines. Such changes in temperature-entropy (T-S) diagram and pressure-heat flux (P-h) diagram for the open Brayton cooling cycle are evident in Figure 5.6.

5.4 Coefficient of Performance for Various Cooling Systems

Regardless of the cooling cycles or refrigerator types, the performance of a cooling system is expressed by a parameter called coefficient of per-

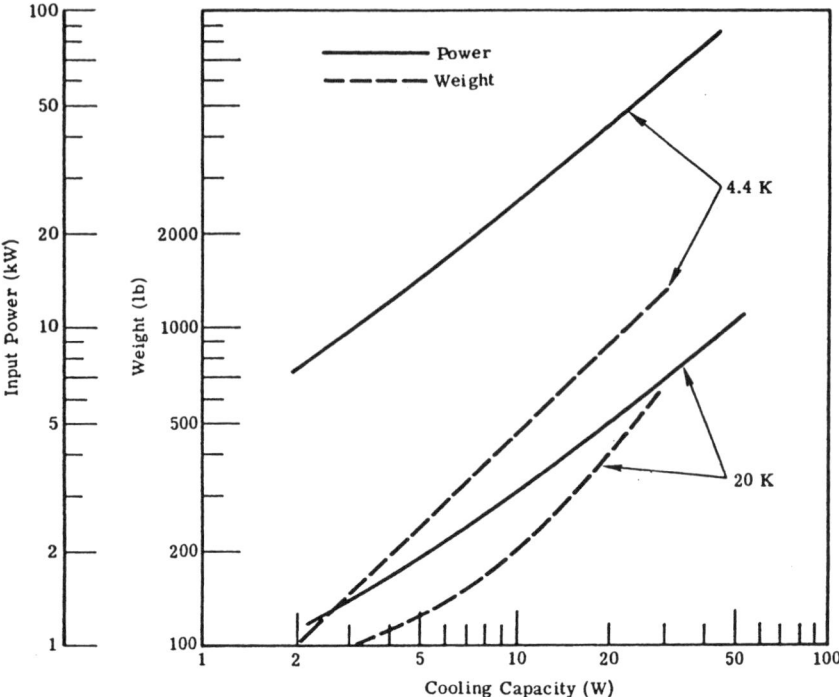

Fig. 5.5 Estimated power and weight parameters for turbo-machinery refrigeration systems operating at two distinct cryogenic temperatures: 4.4 K and 20 K.

formance (COP). COP is defined as the ratio of energy removed from the environment in the form of heat transfer to the energy input to the system. The energy input to the system must include all the energy components involved in the cooling process. In the case of an air-conditioning system, the cooling capacity is generally expressed in tons. One ton is the cooling rate necessary to freeze 1 ton (or 2000 lb) of water in 1 day or within 24 hours. The cooling capacity of 1 ton is equivalent to 12,000 BTU/hour or 3517 watts of electrical power consumption. For a reversible refrigeration cycle or air-conditioning cycle, it removes energy from a low-temperature reservoir at T_L and rejects energy to a reservoir at a higher temperature at T_H. For such an operating cycle, the coefficient of performance can be written as

$$\text{COP} = [T_L/T_H - T_L] \qquad (5.1)$$

5. Performance Requirements for Refrigeration Systems

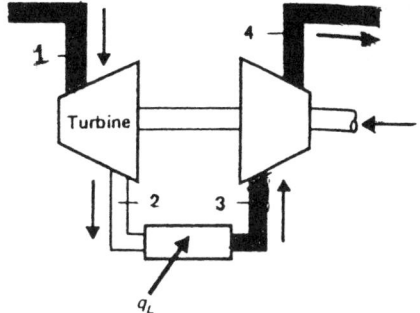

(A) Critical system elements of an open Brayton cooling cycle

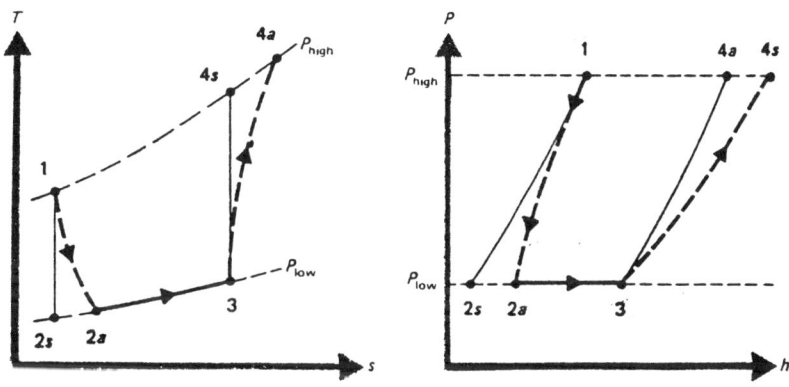

(B) Temperature-entropy diagram   (C) Pressure-heat diagram

Fig. 5.6 Critical system elements (A), temperature-entropy diagram (B), and pressure-heat diagram (C) for an open Brayton cooling cycle. Symbols: T = temperature (K); s = entropy; p = pressure; q = heat energy; and h = heat flux.

Computed values of COP as a function of various operating temperatures are summarized in Table 5.3. It is evident from this table that the COP parameter for a reversible refrigeration cycle improves with lower values of high temperature and higher values of low temperature.

5.4.1 COEFFICIENT OF PERFORMANCE FOR AN IDEAL BRAYTON COOLING CYCLE

Studies performed by the author indicate that an ideal Brayton cooling cycle offers maximum values for the coefficient of performance and hence

Table 5.3 Coefficient of Performance (COP) for a Reversible Refrigeration Cycle.

	$T_H(C)$		
$T_L(C)$	70	80	90
−1	0.0141	0.0123	0.0110
−2	0.0277	0.0244	0.0217
−3	0.0411	0.0369	0.0322
+1	0.0145	0.0126	0.0112
+2	0.0294	0.0256	0.0227
+3	0.0449	0.0389	0.0345

Table 5.4 COP for an Ideal Brayton Cycle Refrigeration System.

(P_H/P_L)	k = 1.5	k = 2.0	k = 2.5	k = 3.0
500	0.1440	0.0468	0.0246	0.0162
400	0.1570	0.0526	0.0387	0.0188
300	0.1760	0.0613	0.0337	0.0229
200	0.2060	0.0761	0.04334	0.0302

is most attractive for industrial and manufacturing applications where performance is of critical importance. The coefficient of performance for an ideal Brayton cycle can be expressed in terms of the temperature and pressure parameters as shown in the T-S and P-h diagrams in Figure 5.6. For isentropic turbine and compressor units, the COP can be written as

$$\text{COP} = [T_3/T_{4,S} - T_3] = \left[1/(P_H/P_L)^{(k-1/k)} - 1\right] \quad (5.2)$$

The temperature terms (T) and pressure terms (P) shown in Figure 5.6 assume low and high values for the temperature-entropy diagram and pressure-heat diagram. The value of parameter k is strictly dependent on the isentropic conditions. Computed values of this COP for an ideal Brayton cycle as a function of high and low operating pressures for various magnitudes of parameter k are summarized in Table 5.4. Close examination of computed values indicates that lower pressure ratios (i.e., high-to-low pressure ratios) offer improved coefficient of performance for an ideal Brayton refrigeration cycle.

5.5 Cryogenic Dewar and Storage Tank Requirements for Various Applcations

The design complexity for a cryogenic Dewar or storage tank varies from application to application and operating environments. The design of a cryogenic storage tank for an aircraft, space vehicle, or satellite communication system must take into account all the operating conditions and critical performance parameters, such as minimum cryogenic temperature, maximum fluid pressure, an economically permissible storage tank configuration (spherical or cylindrical), maximum allowable weight, and a high factor of safety to provide high mechanical integrity under extreme mechanical environments. The safety parameter (F_s) will ultimately determine the cost-effective design capable of providing optimum reliability and risk-free operation.

The design optimization of a high-pressure, cryogenic-gas storage tank must address the critical design issues to minimize the container weight and volume and to maximize the storage pressure with an adequate factor of safety and without incurring excessive pressure-shell weight. If the operating pressure exceeds 1000 atmospheres (1000 psia), gases becomes less compressible so that the volume savings at high operating pressures are significantly diminished. Tradeoff studies must be performed to obtain a cost-effective design with no compromise in reliability or safety.

It is critical to note that the gas compressibility factor for a specific gas will ultimately determine the safe optimum pressure. Weight estimates for high-pressure gas storage tanks depend on the geometrical configuration (cylindrical or spherical) of the tank, the volume or size of the tank, the operating pressure, and the factor of safety requirement. The weight of the storage tanks can be computed using the following empirical mathematical expressions:

$$[W_{\text{cylindrical}}] = [(2 \times 10^{-8})V^{1.07}P^{1.5}F_s^m] \text{pound} \quad (5.3)$$

$$[W_{\text{spherical}}] = [(373 \times 10^{-6})V^{0.86}P^{0.49}F_s^m] \text{pound} \quad (5.4)$$

where W is the weight of the storage tank (lb), V represents the volume of the storage tank (in^3), P indicates the operating pressure (psia), F_s is the factor of safety generally assumed as 1.5 or 2.0, and the exponent m has a value of 1.36 for a cylindrical tank configuration and 0.70 for a spherical tank configuration [2].

Table 5.5 Calculated Weight of a Cylindrical High-Pressure Gas Storage Tank as a Function of Volume (V) and Operating Pressure (P) in Pounds.

Volume (in.^{3})	Pressure (psia)				
	1000	2000	4000	6000	8000
1000	2.63	7.45	21.06	36.69	59.57
5000	14.75	41.74	118.08	216.93	333.94
10,000	30.92	87.46	247.42	454.44	699.74
50,000	172.98	489.58	1384.34	3440.22	5296.5
100,000	363.38	1027.88	2907.57	5340.52	8222.8

Table 5.6 Calculated Weight of a Spherical High-Pressure Gas Storage Tank as a Function of Volume (V) in in.^{3} and Operating Pressure (P) in Pounds per Square Inch (Absolute).

Volume (in.^{3})	Pressure (psia)				
	1000	2000	4000	6000	8000
1000	6.80	9.60	13.41	16.35	18.83
5000	27.13	38.12	53.52	65.28	75.17
10,000	49.25	69.54	97.15	118.49	132.80
50,000	196.56	276.08	387.73	472.92	544.53
100,000	356.77	501.12	703.76	858.38	988.36

It is worthwhile to mention that the exponent "m" used for a cylindrical tank is roughly twice that used in the design of a spherical tank. This is due to the fact that the surface area of a spherical tank configuration is much higher compared to that of a cylindrical tank with comparable physical dimensions. In addition, the large surface area of the spherical tank will experience much lower buckling stress, thereby justifying a lower value of the exponent used for a spherical tank configuration. Computed values of weight for high-pressure gas storage tanks are summarized in Table 5.5 and Table 5.6 for cylindrical and spherical tank configurations, respectively, assuming a factor of safety of 2.0 in each case.

Weight estimates as a function of tank volume and operating gas pressure for a spherical high-pressure gas storage tank can be instantly

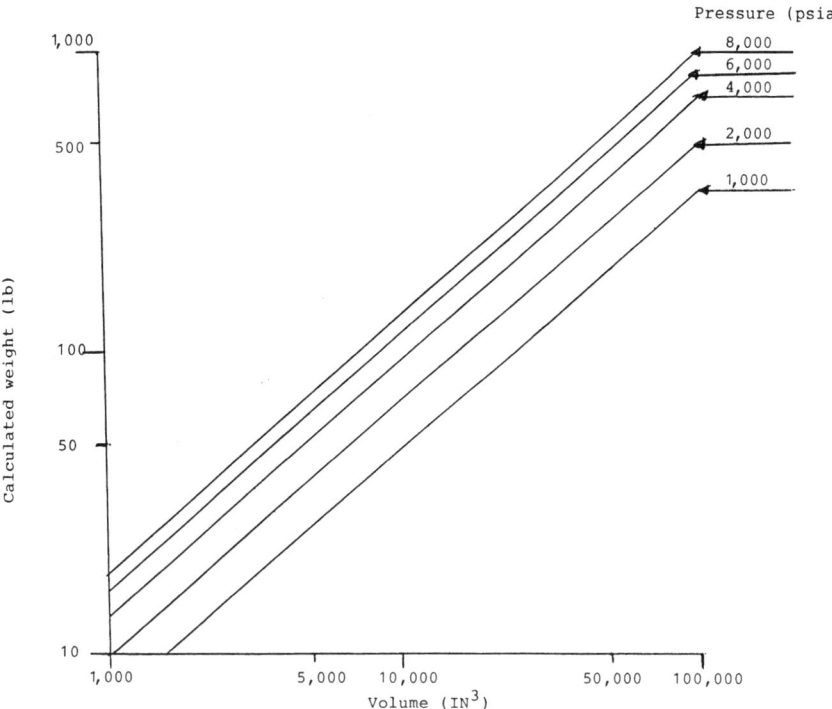

Fig. 5.7 Calculated weight estimates for a spherical high-pressure gas storage tank as a function of operating pressure and tank volume.

obtained by looking at the curves shown in Figure 5.7. Note that the weight estimate curves for a cylindrical high-pressure gas storage tank will have the same trends. These curves, based on least-squares correlation, offer first-order storage tank weight estimates for the aircraft and space configuration systems. These weight estimates include only the basic storage tank structural weight and do not include the weight of the associated hardware such as regulating valves, pressure relief valves, emergency shut-off valves, and transfer lines. Transfer lines are composed of flexible, vacuum-based, superinsulated lines that come in several sizes and strengths and are used to provide effective transfer of cryogenic liquids with minimum heat loss. Heat-loss factors are the critical design parameters of the transfer lines.

Published literature [1] indicates that the accuracy of the calculated weight of high-pressure gas storage tanks made from high-performance cryogenic materials could vary from +/−10% to +/−15%, depending on the size of the tank and operating pressure. However, based on preliminary calculations, higher accuracy is predicted at operating pressures greater than 3000 psia and storage volume exceeding about 10,000 in.³

High performance cryogenic Dewars or storage tanks generally use vacuum, multilayered insulation (MLI) blankets to minimize heat loss under extreme temperature-fluctuating environments. Rigid-radiation shields provided are cooled by the vented vapor to minimize the heat-leakage necessary for extended operations in space. Advanced technologies used in the design of liquid oxygen and hydrogen storage systems for the Gemini and Apollo space programs allowed the storage of cryogenic fluids such as hydrogen and helium for periods exceeding a year in space.

5.6 Storage Tank Requirements for Space and Missile Applications

For space and missile applications, cryogenic fluids require special precautions during the transfer and storage phases. In such applications, cryogenic fluids are required to be stored at lower pressures as liquids in equilibrium state with their vapors known as subcritical environments or at higher operating pressures and temperatures as supercritical, homogeneous cryogenic fluids. However, for laboratory or airborne applications, the cryogenic fluid is stored in the liquid two-phase state because of its compact size, simplified design, logistic advantage, and reduced storage weight. Note that in a space-based storage system for cryogenic liquids, the absence of gravity or gravitational forces and acceleration from orientation forces prevents the use of a standard two-phase system, because random orientation of the liquid phase under weightless environments prevents continuous communication between the liquid phase and the supply port of the storage Dewar. Storage of cryogenic liquids in space is accomplished through pressurization of the cryogen to the supercritical pressure known as a single-phase state. Under space environments, the absence of gravitational forces or acceleration does not have any adverse impact on the delivery of the cryogenic fluid, because the supply port of the Dewar is in direct communication with the homogeneous fluid at all times. Cryogenic storage Dewars provide the desired cryogenic operating temperatures when appropriately attached to a specific system operating in space environments.

5. Performance Requirements for Refrigeration Systems

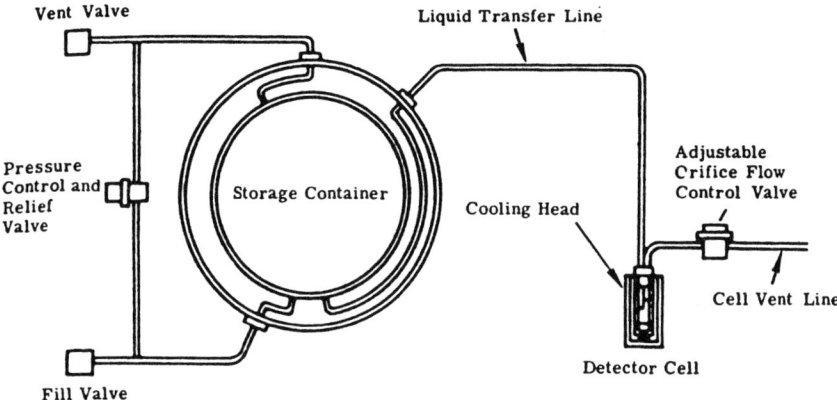

Fig. 5.8 Critical elements of an open-cycle, expandable cooling system using the liquid-feed concept [3].

5.6.1 LIQUID-FEED REQUIREMENTS FOR STORAGE SYSTEMS

The classification of cryogenic storage systems is strictly based on cryogen feed design configurations. Two basic types of cryogenic storage systems include a direct-contact design configuration and a liquid-feed design configuration [3]. Studies performed by the author on feed configurations indicate that a liquid-feed cooling system (Fig. 5.8) offers a cost-effective design with improved reliability and enhanced efficiency. Critical elements of a liquid-feed system include a vent valve, a pressure control and relief valve, a cooling head containing the detector cell, a filling valve, a liquid transfer line, a liquid transfer valve, a cell vent line, and an adjustable orifice flow control valve designed for proper regulation of the flow as illustrated in Figure 5.8.

5.6.2 TRANSFER LINE REQUIREMENTS

Superinsulated and flexible transfer lines are available in a variety of shapes, sizes, and strengths. Transfer lines are designed to provide a reliable and unrestricted flow of cryogenic fluids with minimum heat loss. Transfer line must be wrapped with high-performance insulation blankets if minimum heat loss is of critical importance. Critical performance parameters [3] on flexible transfer lines with associated heat-loss factors for liquid helium transfer at room temperature (300 K) are summarized in

Table 5.7 Critical Performance and Physical Parameters of Selected Vacuum Insulated, Flexible Cryogenic Transfer Lines Carrying Liquid Helium @ 300 K.

Nominal line Size (in.)	Outer hose ID (in.)	Inner hose ID (in.)	Heat loss per foot (BTU/hour)
0.250	2.250	0.125	0.30
0.500	2.852	0.500	0.40
1	4.100	1.000	0.56
2	5.352	2.000	0.73
3	6.500	3.000	0.91
4	8.25	4.000	1.13

Table 5.7 with emphasis on heat loss. As the nominal size of the transfer line increases, heat loss increases because of larger exposed surface area.

5.7 Operating Pressure and Temperature Requirements for Storage of Liquefied Gases

In the case of tanks of liquefied gases, the operating temperature range could vary from 2 to 300 K involving argon and nitrogen gases as illustrated in Figure 5.3. It is important to remember that a change in operating pressure permits temperature variations in a liquefied gas medium to provide cryogenic cooling from the triple point to the critical point. Storage requirements for liquefied gases are quite different than those for cryogenic fluids, because compressibility properties for gases are different than those for cryogenic fluids. Cooling capacity for both argon and nitrogen increases rapidly in the beginning with the increase in operating pressure as shown in Figure 5.3. However, after reaching their respective peak values, the cooling capacity drops with a further increase in pressure. Thus, any increase will result in higher storage costs.

5.8 Cooling System Configurations Using Various Cooling Agents or Cryogens

Performance characteristics and operating requirements for various cooling system configurations must be carefully examined before they are selected for a specific application. Critical operational and logistic problems associated with a specific cryogen cooling system configuration must be evaluated to meet specific performance requirements. Cryogens or refrigerants come in three distinct forms: gaseous, liquid, and solid. Important properties of these cryogens will be briefly described. Design configuration complexity depends on the cryogen used to meet overall system requirements, low maintenance costs, minimum cooling time, and high reliability.

5.8.1 CHARACTERISTICS OF VARIOUS CRYOGENS

Liquid and gaseous cryogens are widely known as expendable coolants that use the heat of sublimation or vaporization. The operating temperature limits of such coolants are of critical importance. Note that the operating temperature limits for a coolant are based on a minimum temperature as defined by the solid phase at 0.1 mm of mercury pressure and on a maximum temperature defined by the critical point associated with that particular coolant. Typical operating temperature limits or ranges for the expendable coolants (i.e., liquid and gaseous coolants) are shown in Figure 5.9. Cooling temperature ranges can be satisfied by selecting an appropriate coolant or cryogen capable of meeting the environmental and performance requirements. The cooling capacity of a cryogen depends on the thermodynamic properties of the selected expendable refrigerant or cryogen. The thermodynamic properties of selected expendable cryogens are summarized in Table 5.8.

Table 5.8 Thermodynamic Properties of Widely Used Expendable Cryogens [3].

Cryogen name	Boiling point @ 1 atm (K)	Melting point @ 1 atm (K)	Gas density @ NTP (g/l)	Liquid density @ BP (g/ml)	Critical temperature, K
Helium	4.2	2.0	0.173	0.125	5.2
Hydrogen	20.39	13.98	0.090	0.071	33.2
Nitrogen	77.4	63.4	1.250	0.808	126.1
Argon	87.4	83.6	1.780	1.391	150.8

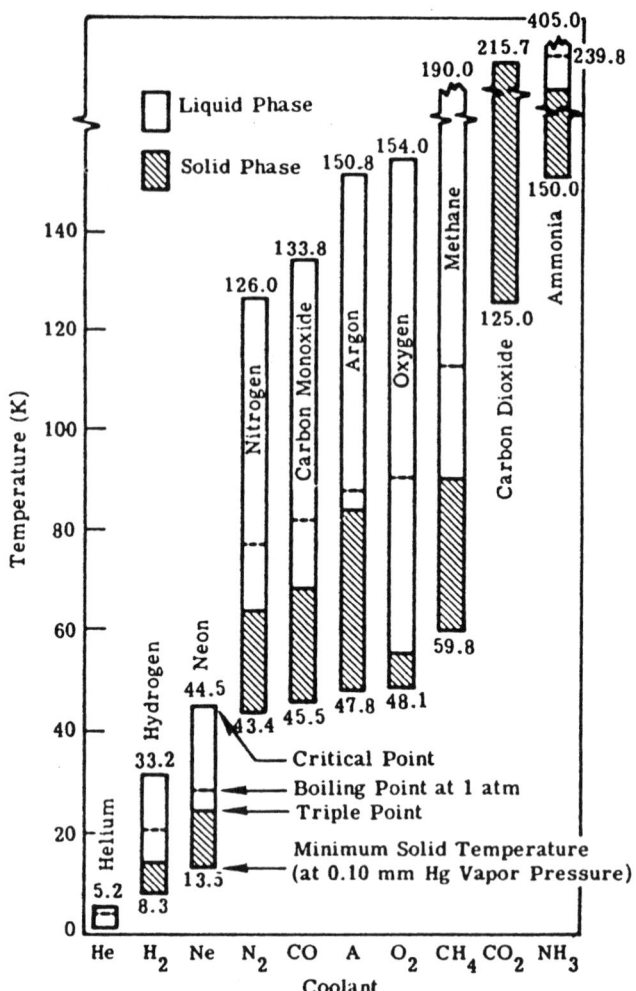

Fig. 5.9 Typical operating temperature ranges and boiling points for selected expendable coolants showing their solid and liquid phases [2].

5.8.2 SOLID CRYOGENS

A cooling system based on the sublimation of a solid cryogen into a high vacuum space is most desirable, because it avoids several operational and logistic problems associated with either the subcritical or the supercritical storage of coolants. In some applications, a specific solid cryogen may

Table 5.9 Properties of Selected Solid Cryogens [3].

Cryogen	Heat of sublimation (BTU/lb)	Density of solid @ MP (lb/ft³)	Operating temperature (K)@	
			0.1 mm of Hg	triple point
Ammonia	739	51.3	150.2	195.4
Argon	79.8	7.0	47.8	83.6
Hydrogen	218.5	5.02	8.30	14.0
Nitrogen	96.6	63.8	43.4	63.4

be of significance. A solid-cryogen-storage cooling system consists of a solidified cryogenic substance, a vent-gas path to space, an insulated cryogen container, and a conduction path from the coolant to the system under consideration. Critical properties of a solid cryogen include the peak pressure of the vent gas, operating time, heat load, and cooling capacity. The operating temperature depends on the coolant choice and maintaining the right back pressure of the vent gas for satisfactory operation of the system. However, the operating time requirement is contingent on the amount of coolant and the heat-load factor. Furthermore, the cooling capacity is dependent on the heat of sublimation, which is equal to the sum of the heat of fusion and the heat of vaporization.

A solid-cryogen system has several operational advantages over a liquid-cryogen system, including independence from an operating altitude, a high cooling capacity depending on the type of solid-cryogen used, a high-density storage configuration, and a low-temperature solid phase capable of providing higher sensitivity in some infrared (IR)-detection systems. The operation of a solid-cryogen system is based on the interrelation of the temperature and pressure of a solid equilibrium with its pressure. During the cooling process, the addition of heat sublimes the solid coolant and increases higher vapor pressure, leading to a temperature increase. However, the right pressure and temperature must be maintained at constant levels by venting the vapor to space at a specified pressure level. Generally, the coolers are designed without valves; nevertheless, the vent gas ducting is provided to maintain a required pressure. Normal operating temperature ranges vary between 8.3 and 13.98 K [4] for hydrogen, 43.4 and 63.4 K for nitrogen, and 47.8 and 83.6 K for argon at a pressure of 0.1 mm of mercury and triple point, respectively. Important properties of most widely used solid cryogen are summarized in Table 5.9.

5.8.3 TECHNIQUES TO REDUCE HEAT LEAK AND WEIGHT OF CRYOGEN

Minimization of heat leak must be given serious consideration, which is dependent on the Dewar technology incorporating multilayer insulation (MLI), low-conducting support materials, and unique isolation techniques. Good thermal contact must be maintained using expanded metal foams and wire-mesh heat exchangers built into the solid-cryogen tank. The heat leak into the cryogen is through the MLI medium and the heat absorbed by the cryogen is in the product of the heat leak and elapsed time.

The weight of the cryogen required to meet a specific mission goal in space is contingent on the temperature to be maintained, heat leakage, duration of the mission, and the type of cryogen used in the tank. Exact weight determination is very difficult, because of variable atmospheric characteristics in space environments. The studies performed by the author indicate that the weight of a cryogen required to meet a specific operating temperature is dependent on the type of cryogen used, cryogen properties, the mission time requirement, and unstable operating environments. The studies further indicate that the weight of a cryogen required to meet a certain cryogenic temperature within a specified time frame is inversely proportional to the square of cryogen density, inversely proportional to the cube of the latent heat of the cryogen, directly proportional to the third power of the product of effective emissivity and elapsed time, and directly proportional to the difference of 12th power of outer shell temperature (K) and inner shell temperature.

5.9 Performance Comparison of Various Cryogenic Coolers

The design concepts and performance capabilities of various cryogenic cooling systems will be described in terms of cooler type, cooling temperature requirements, cooling capacity, the cryogen used, elapsed time, advantages, and disadvantages. An open-cycle extended cryogenic system includes a liquid cooler system (subcritical), a single-phase storage cooler system (supercritical), a gas storage cooling system incorporating the J-T-expansion valve mechanism, and a solid-cryogen storage cooling system. The mechanical refrigeration system configurations include Stirling coolers, VM coolers, G-M coolers, closed-cycle J-T coolers, and

Table 5.10 Performance Characteristics of Various Cryogenic Systems.

System type	Cooler type	Temperature range (K)	Cooling capacity (W)	Advantages	Disadvantages
Open-cycle expendable	Solid-cryogen storage	8–150	0.010–1	Reliable low density	Complex costly
	Gas-cryogen with J-T valve	4.2–87.4	20 (max)	Remote operation	High pressure High gas reqt.
	Single-phase (supercritical)	5.2–126	Unlimited	Flexible	Excessive Dewar weight
	Liquid Storage (subcritical)	4.2	Unlimited	Reliable	Complex
Mechanical refrigerators	Stirling cycle G-M cycle	10–77 10–77	0.010–15 0.005–15	Compact airborne Best for airborne	Limited life High-power consumption
	Claude cycle	77	0.500–5	No logistic problems	Limited life High power
	Turbine-based Brayton-cycle	4.2–77	10–100	Max. life	High cost High power

Brayton-cycle coolers. Critical performance parameters as well as the advantages and disadvantages of various cryogenic cooling systems [4] are summarized in Table 5.10. Major emphasis must be placed on cost, complexity, and reliability.

It is important to distinguish between the mode of operations of open-cycle and closed-cycle refrigeration systems. Open-cycle expendable systems use cryogens in either the subcritical or the supercritical liquid form. However, closed-cycle, mechanical refrigerator systems provide cooling at low temperatures but reject heat at relatively high temperatures. Both the Sterling-cycle and G-M-cycle coolers are best suited for airborne or space applications because of high reliability.

5.10 Summary

This chapter described refrigeration systems with moderate- and high-capacity cooling, with an emphasis on cooling capacity, cryogen consumption rate, cooldown time, and reliability. Performance requirements and design parameters of high-pressure gas refrigeration systems were summarized, identifying the critical operating parameters. Brayton-cycle, Claude-cycle, G-M-cycle, and reverse-Brayton-cycle refrigeration systems were described, with an emphasis on the cooling load, working fluid, and cooldown time. Calculated values of the coefficient of performance (COP) parameter for various cooling systems were provided. Design requirements for cryogenic storage tanks were discussed in great detail. Computed values of weight estimates for high-performance cryogenic storage tanks as a function of operating pressure and container size were provided. Important characteristics of gas, liquid, and solid cryogens were summarized, with a major emphasis on operating temperature. Physical parameters and heat-loss factors for vacuum-insulated cryogen transfer lines were summarized, with an emphasis on heat-loss sources. The performance characteristics of various cryogenic cooling systems, including open-cycle expanded refrigerators and mechanical refrigeration systems, were summarized, with a major emphasis on temperature range, cooling capacity, reliability, and design complexity.

References

1. A. R. Jha. *Superconductor technology: Applications to microwave, electro-optics, electrical machines, and propulsion systems.* New York: John Wiley & Sons, 1998, p. 276.
2. Wolfe and Zissis, *The infrared handbook.* 1978, effective spectral band, pp. 14–15.
3. Wolfe and Zissis, *The infrared handbook.* Modulation Transfer, pp. 21–22.
4. Jha, *Superconductor technology.* pp. 279–281.

Chapter 6 | Cryocooler and Microcooler Requirements Best Suited for Scientific Research, Military, and Space Applications

6.0 Introduction

This chapter focuses on the performance requirements for cryocoolers and microcoolers best suited for military, space, medical, scientific research, and high-resolution system applications. The performance capabilities of various cryocoolers widely used in microwave, MM-wave, infrared (IR), and electrooptic (EO) sensors are summarized, with an emphasis on reliability, maintenance requirements, and cooling capacity. Cryocoolers for both passive and active devices and systems are briefly described. Significant improvements in the performance of cryogenically cooled microwave components, photonic devices, infrared detectors, focal planar arrays (FPAs), high-resolution CCD-based cameras, missile seeker receivers, high-resolution imaging sensors, and complex military systems are identified. Performance requirements for microcoolers best suited for space-based systems are summarized with major emphasis on reliability, power consumption, cooling temperature, cooling capacity, and weight.

A cost-effective cooling scheme must be selected that can be integrated into cryogenic sensors. This scheme must meet the operating temperature requirements with a minimum coolant amount and high cooling efficiency, and it must do so with minimal cost and complexity. The weight, size, cost, and design complexity depend on the cooling capacity and the operating temperature to be achieved in a specified time frame. The chapter briefly discusses the regular maintenance and servicing requirements necessary to ensure that a cooling system is safe, steady, and reliable. Maintenance requirements for microcoolers best suited for scientific research, space, and military applications are summarized with an emphasis on reliability and maintenance-free operation over a long duration.

6.1 Cryocooler Requirements for Various Applications

The cryogenic cooling of IR detectors, CCD-based FPAs, high-performance detection sensors, missile electronics and optics is absolutely necessary if reliable, stable, and optimum performance are the principal requirements. As stated earlier, weight, size, cost, and complexity are contingent on cooling capacity and cryogenic operating temperature to be achieved in the shortest duration. The regular maintenance and servicing of cryocoolers and microcoolers are required to ensure uninterrupted reliable operation of the cryogenically cooled system or device. Commercial, industrial, military, and space-based cryogenically cooled sensors must address the following critical design issues associated with cryocoolers:

- The decision to deploy a specific cryocooler or microcooler must be based on comprehensive design analysis to meet specific performance requirements.
- Selection of a cryocooler design configuration must consider the most critical requirements, such as ease of operation, minimum coolant consumption, and less-frequent maintenance servicing.
- Effective integration of the selected cryogenic scheme with minimum cost and complexity but with no compromise on reliability must be given serious consideration.
- Maintenance procedures in the field with minimum cost and supervision must be addressed.

Studies performed by the author indicate that these design features have been implemented in few commercial IR sensor applications. However, these design features have been integrated in high-performance military IR surveillance and target acquisition systems, space reconnaissance sensors, and magnetic resonance imaging (MRI) diagnostic equipment.

Closed-cycle and open-cycle cryogenic refrigeration systems are now available with various cooling capacities. The selection of a cryocooler for a specific application depends on cooling capacity, operating temperature, procurement, cost and maintenance, and servicing requirements. Data on the history of cryocooler development and a survey of currently operating cryogenic systems for commercial, military, and space applications are summarized in Figure 6.1. Performance capabilities of both high-temperature (77 K) and low-temperature (4.2 K) cryocoolers are briefly described, identifying the type of cryocooler and its user.

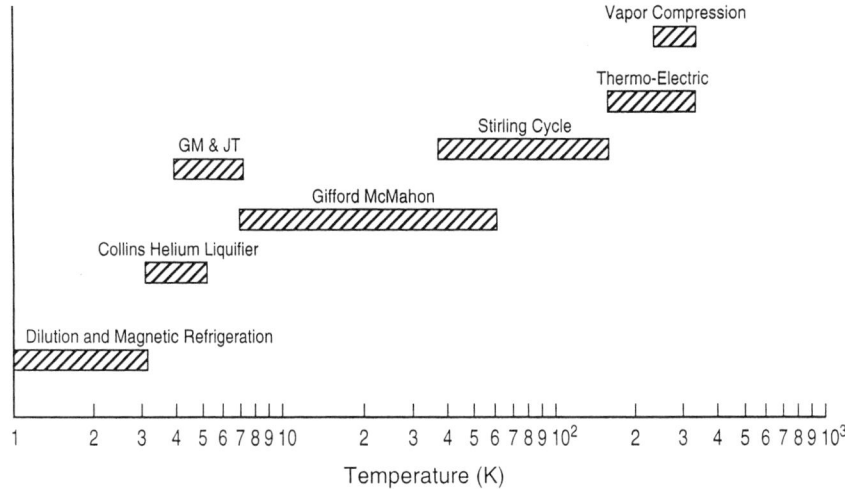

Fig. 6.1 Temperature ranges for commercial refrigerators [1].

Closed-cycle and open-cycle cryogenic refrigeration systems will be described in greater detail with major emphasis on cooling capacity, operating cost, and maintenance issues. A survey of commercially available cryocoolers indicates that most systems suffer from maintenance problems and coolant leakage when operated over extended durations. For tactical missile applications and for space mission operations, the maintenance of cryocoolers is sometime out of the question. However, maintenance issues for commercial, medical, and scientific research cryogenic refrigerator systems must be given serious consideration if reliable operation is the principal requirement.

6.1.1 MAINTENANCE REQUIREMENTS FOR CRYOCOOLERS

Maintenance cost and interval requirements must be seriously considered during the selection phase of a cryocooler for a specific application. It is important to mention that the maintenance interval criterion can be determined only from the maintenance data on the cryocoolers operating in the field and the experience of the maintenance personnel. According to the General Electric maintenance personnel [1], a cryocooler operating at a cryogenic temperature of around 4.2 K requires more elaborate and more frequent maintenance servicing than a cryocooler operating at a cryogenic temperature of 77 K or higher. This means maintenance costs will be higher for cryocoolers using liquid helium coolants, because complex helium purification devices are required to maintain clean working fluid in the system at all times. Liquid helium cryocoolers generally run smoothly in the laboratory under controlled environments, but when liquifiers are incorporated in the complex system configuration such as MRI, they failed to perform with high reliability. However, cryocoolers operating at higher temperatures have more successfully met their stated reliability goals and maintenance-servicing requirements.

6.2 Performance Parameters for Various Cryocoolers

Critical performance parameters for various commercial, industrial, scientific research, and space applications are summarized with emphasis on operating temperature and coolant capacity. Operating temperatures for various cryocoolers are shown in Figure 6.1. For ultra low temperature applications, three distinct cryocoolers — Collins helium liquifiers (CHLs), Gifford-McMahon (G-M) crycoolers, and G-M/J-T cryocoolers with Joule-Thomson valves — are widely deployed. Brief technical descriptions for various crycoolers are provided for readers not familiar with cryogenic technology.

6.2.1 DILUTION-MAGNETIC CRYOCOOLER

These cryocoolers are widely used in applications requiring very low temperatures, where a small sample needs to be cooled below a cryogenic temperature of 3 K. They are best suited for scientific research applications in laboratory-controlled environments.

6.2.2 COLLINS-HELIUM LIQUIFIER (CHL)-BASED CRYOCOOLER

CHL cryocoolers belong to a broad class of refrigeration systems based on the Claude-cycle operating principle. This system employs a combination of two or more expansion devices or engines depending on the cooling temperature requirements. Recuperative heat exchangers or Joule-Thomson (J-T) expansion valves are used to supply cryogenic fluid at a

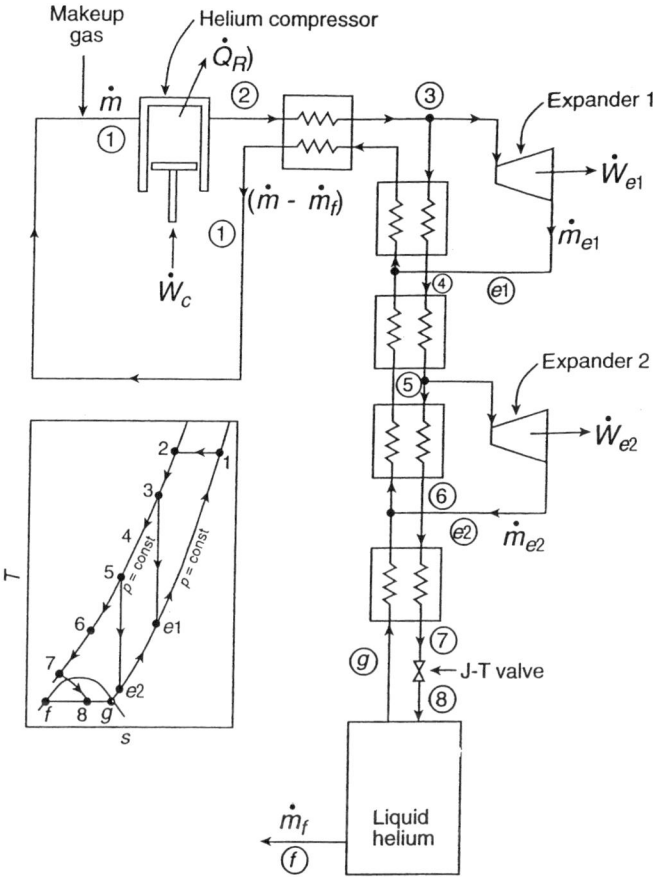

Fig. 6.2 Schematic diagram for a Collins-cycle liquid helium refrigerator [1].

temperature of 4.2 K. Critical elements of a CHL cryogenic cooler are illustrated in Figure 6.2. These cooling systems are available with cooling capacities ranging from 15 to 90 watts of refrigeration consuming 5 to 75 liters per hour of liquid helium at 4.2 K operation. This particular closed-cycle refrigerator has been widely used in high-energy accelerators, DC superconducting fusion magnets, magnetohydrodynamic (MHD) systems, and magnetically levitated trains. Operating temperatures and users of various cryocoolers are identified in Table 6.1.

Table 6.1 Operating Temperatures and Users of Various Cryocoolers for Various Applications.

Application	Temperature (K)	Refrigerator	User
1960s			
Infrared	80	Stirling	U.S. military
Hydrogen bubble chamber	20	GM	High-energy research
Microwave amplifiers	4.5	GM + JT	NASA
1970s			
High-energy accelerators	4.5	Collins	DOE
DC superconducting motor	4.5	Collins	U.S. Navy
Fusion	4.5	Collins	DOE
1980s			
Fusion magnet test facility	4.2	Collins	DOE
30-MJ SMES	4.5	Collins	BPA
Magnetohydrodynamics	4.5	Collins	Russia
Magnetically levitated train	4.5	Collins	Japan
Magnetic resonance imaging	4.2	GM	Commercial
Cryovacuum pumping	20	GM	Commercial
1990s			
SQUID	8.5	Stirling	NBS
SSC	2.0	Supercritical	DOE
MRI	10	GM	Commercial
Magnetically levitated train	—	—	DOT
SMES	—	—	Industry
Infrared	65	Stirling	U.S. military
Infrared	10	Stirling	NASA
High Tc	20–30	—	DOE/military

6.2.3 GIFFORD-MCMAHON (G-M) CRYOGENIC REFRIGERATOR

This particular cryogenic cooler has been most successful in operations requiring cryogenic temperatures in the 6- to 8-K range. These cryocoolers are widely used in cryogenic vacuum pumps and MRI systems, where minimum helium boiloff through reduction of thermal losses in the cryostat is the principal operating requirement. A G-M closed-cycle, two-stage

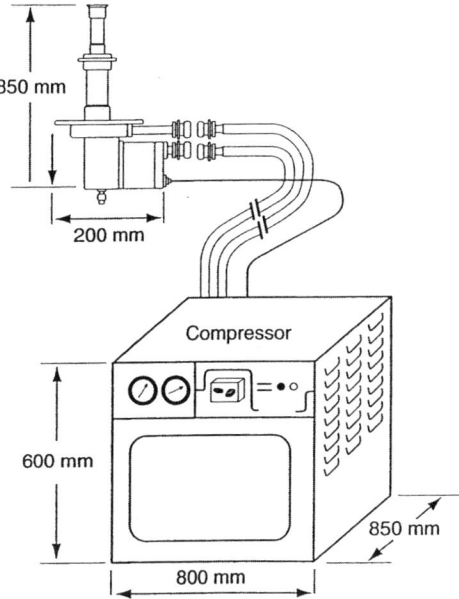

Fig. 6.3 Gifford-McMahon cycle, two-stage refrigerator for MRI applications [1].

refrigerator shown in Figure 6.3 is best suited for MRI applications [1] and is most cost-effective because of minimum helium loss without any adverse effect on system performance. There is a tremendous demand for G-M refrigerators because of uninterrupted, maintenance-free operation over extended periods. According to market published reports, more than 1800 units are manufactured and distributed worldwide, mostly for MRI applications.

6.2.4 G-M/J-T REFRIGERATOR SYSTEM

The G-M refrigerator system, when integrated with J-T valves, offers the most cost-effective way to provide cooling specifically at a cryogenic temperature of 4.2 K. From 1980 to 1995, both the superconducting magnets and cryopumps made exclusive use of two-stage G-M refrigeration systems to provide effective conductive cooling instead of immersion cooling. Conduction cooling is mostly free from logistic and major reliability problems. In the case of an MRI system, the superconducting magnet is immersed into the liquid helium bath at an operating temperature of 4.2 K. The G-M cryocooler cools the magnetic radiation shield that is used to reduce the helium boiloff to a very low level, and thus the coolant level can be maintained with infrequent helium delivery as small as once a year. However, the cryopumps generally use G-M refrigerators with conductive cooling to cool the cryopumping surface to about 20 K temperature. Accelerated research and development activities on conductively cooled magnets for MRI applications and deployment of improved high-temperature superconducting materials will unquestionably improve the performance of future MRI systems in terms of efficiency and reliability. As stated earlier, an MRI conductively cooled magnet uses a two-stage G-M refrigerator to cool both the magnet and the radiation shield, thereby requiring only a few watts of costly refrigerant or cooling agent.

Superconducting quantum interference device (SQUID) technology is widely used in military, medical, and scientific applications. SQUID technology can be easily integrated with closed-cycle G-M or Stirling cryocoolers because of its unique design compatibility. It is interesting to note that from 1960 to 1980, G-M/J-T crycoolers were widely used in low-noise microwave amplifiers, whereas the Stirling coolers were used in a number of military IR and EO systems with maintenance interval requirements of better than 4000 hours. MRI and cryopumping systems generally require a two-stage cooling scheme, one over 80 K to 20 K temperature range with 40 to 50 watts of cooling power (first stage) and the other at 4.2 K cryogenic temperature with 4 watts of cooling power (second stage), respectively. These cryocoolers normally require only yearly maintenance service because of lowest helium loss and minimum downtime. This type of cryocooler is best suited for space-based sensors, where high reliability and minimal cost are the critical design requirements.

6.2.5 STIRLING-CYCLE CRYOCOOLER

The latest research on bearing and seal technologies and the availability of high-quality lubrication has led to an optimum design of a Stirling-cycle (SC) cryocooler that has demonstrated uninterrupted and unattended operation of several years. Potential applications of this cryocooler include space sensors and complex, unattended military systems in secluded areas, where continuous operation with high reliability is of critical importance. SC cryocoolers are best suited for applications where compact size, minimum weight, and high reliability over extended periods are the principal requirements. Furthermore, these coolers are widely used in cryogenic systems requiring small cooling capacities ranging from 1 to 5 watts over 60 to 80 K operations. SC cryocoolers with unique linear drive mechanisms instead of old rotary drive mechanisms are currently in great demand for space and military applications to meet the requirement for high reliability over extended durations.

The latest SC cryocooler design uses an orifice pulse tube that eliminates the need of the moving displacer from the expander unit, thereby leaving only one moving component in the compressor assembly. The linear drive mechanism can be operated down to a temperature of 60 K with no compromise in cooler reliability. Multistage versions of this cooler are currently in the development phase and will be available in the near future.

6.2.6 SELF-REGULATED J-T CRYOCOOLER

This particular cryocooler is essentially a G-M cryocooler design incorporating J-T-valve technology that offers helium liquid cooling capability at 4.2 K or lower with a minimal loss of refrigerant and with no compromise in the cooler reliability or performance. The self-regulation capability [2] of the J-T valve offers significant improvement in the overall performance of the cryocooler.

6.2.7 CLOSED-CYCLE, SPLIT-TYPE (CCST) STIRLING CRYOCOOLER

This particular Stirling cryocooler is sometimes referred to as a viable alternative to a liquid helium cooling system. The CCST version uses helium gas as the working fluid. The compact CCST helium gas cryocooler shown in Figure 6.4 is best suited for SQUID sensor applications and has only two moving parts: a piston and a displacer. During the compression-operating mode, heat is produced, which is rejected at the ambient temperature T_1. During the expansion mode, the cryogenic fluid is expanded, generating a flow of heat into the fluid at temperature T_2. The fluid picks up the heat from the walls of the annular gap as it travels from temperature T_2 to T_1. For a steady-state operation, the heat flux Q_a must be equal to Q_b.

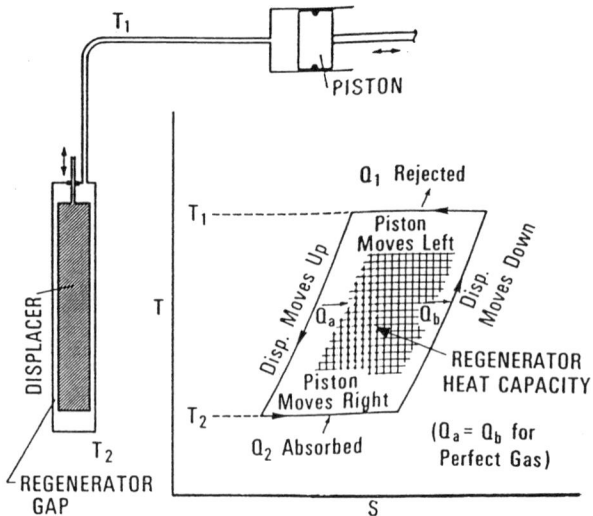

Fig. 6.4 Compact, closed-cycle, helium gas, split-type Stirling cryocooler.

The enthalpy change in the heat quantity or capacity (Q) at two distinct temperatures depends on operating pressure. Under these circumstances, the expansion cannot be isothermal, which is a serious limitation because the heat capacity of the walls at cold ends becomes insufficient. This drawback can be overcome by using advanced materials with large heat capacities and high thermal conductivity in a regenerative heat exchanger at low temperatures. The materials selected for the construction of the heat exchanger must provide adequate protection against electromagnetic interference (EMI) and electronic noise entering the SQUID sensor. Specific details on a single-stage displacer-refrigerator system using advanced nonmagnetic materials [3] are shown in Figure 6.5. This particular cooler machine operates at 1 Hz and obtains a cryogenic temperature of 50 K within less than 4 hours, which can be maintained for about two days on a cylinder of helium. Based on the single-stage Stirlng cryocooler design, the three-stage, closed-cycle Stirling crycooler [3] shown in Figure 6.6 was developed, and it incorporates advanced materials and components — namely, nylon displacer, epoxy-glass cylinder, and aluminized-plastic radiation shield. This cryocooler can maintain an operating temperature well below 16 K when operated at a frequency of 1 Hz. By eliminating the dead space produced by the difference in the coefficient of expansion of various materials used and by optimization of the phase angle between

6. Cryocooler and Microcooler Requirements

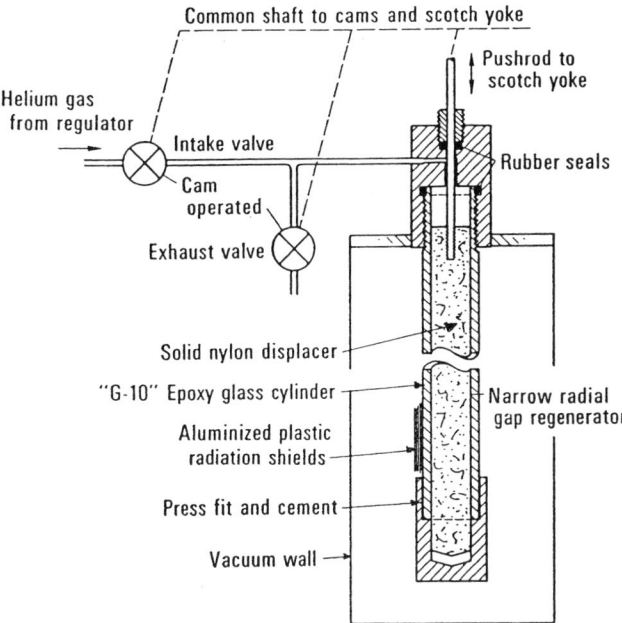

Fig. 6.5 Single-stage, displacer-regenerator cryocooler using advanced nonmagnetic materials [3].

the piston and displacer, this particular cryocooler is able to maintain a temperature below 13 K over a continuous duration of as long as 5 weeks. This cryocooler has demonstrated nonstop operation exceeding 5000 hours with no sign of wear and tear, while maintaining the operating temperature between 12.5 and 13.0 K at the cold end with no heat load. The outer radiation shield temperature is maintained at 120 K, and the inner radiation shield temperature is maintained at 40 K. When operated at a frequency of 1 Hz, the mechanical power required during the compression cycle is about 15 watts based on an isothermal process. Since a considerable fraction of work is returned during the expansion cycle, the net power consumption is much less than 10 watts.

It is important to mention that this particular low-power cryocooler can maintain an operating temperature well below 15 K over a very long period with 10 to 15 mW of heat load. This three-stage cryocooler distributes the refrigeration capacity in such a way that the bulk of the heat input due to conduction and radiation is pumped at relatively higher

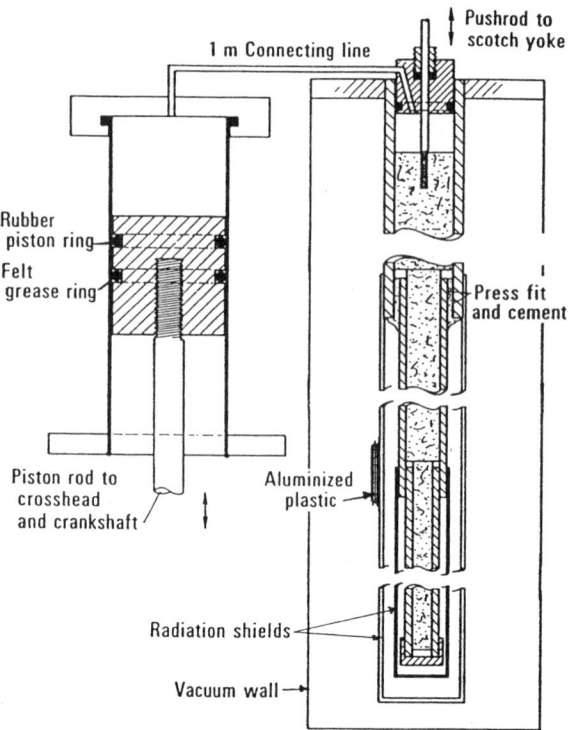

Fig. 6.6 Three-stage, closed-cycle, split-type Stirling cryocooler using advanced non-magnetic materials [3].

temperatures, thereby leaving little heat for the low temperature. The optimum distribution of the refrigeration capacity as a function of temperature is important in a cryocooler design. Note that a small liquid-helium cryostat with an evaporation rate of 1 liter per day, which corresponds to 700 liters of gas at room temperature (300 K), can support a total heat load of 29 mW at a cryogenic temperature of 4.2 K and still has the capability to absorb an additional load of 7.6 mW per degree K due to the heat capacity of the vapor. This particular cryocooler is best suited for space research applications because of its minimum power consumption. Note that the cost of liquid helium will increase as the supply diminishes, whereas the cost of the cryocooler will decrease significantly if the microcoolers are manufactured on a mass scale using cost-effective production techniques.

6.3 Cooling Schemes Used by Various Cooling Systems

The selection of a cooling scheme depends on the cooling capacity, the time required to attain a specified cryogenic temperature, and the type of cryogen or refrigerant used by the cryocooler. Potential characteristics of various cryogens are summarized in Table 6.2.

6.3.1 CHOICE OF COOLING SCHEMES

The choice between immersion cooling and conduction cooling is a fundamental design issue based on the operating requirements of the cryocooler. In the case of immersion cooling using liquid helium, one can use a Collins liquifier for a closed-cycle operation or for an open-cycle operation with a supply of liquid helium as needed. For a low-thermal loss cryogenic operation in the order of 5 watts or less, G-M/J-T cryocoolers using liquid helium as a working fluid are widely used. Under transient state operation during cooldown phase, the helium boiloff is allowed to escape from the cryostat, and the cryostat is provided with new supply of liquid helium. Frequent maintenance is required irrespective of the cooling scheme or cryogen used to keep the system operating with high efficiency and reliability over extended periods. Because of frequent maintenance, immersion cooling is relatively costly. On the other hand, conduction cooling eliminates the need for frequent liquid helium refills, thereby yielding a most cost-effective operation.

Most commercial cryocoolers have small cooling capacities (less than 5 watts) and provide a single point of cooling. The use of such coolers requires complete familiarity with termal load requirements, maximum allowable operating temperatures, and optimum thermal design to mini-

Table 6.2 Potential Characteristics of Most Common Cryogens.

Cryogen	Boiling temperature @ 1 atm (K)	Melting temperature (K)	Rel. vaporization	Power (W/W)
Helium	4.2	2.0	1*	1000
Neon	27.2	24.5	41	140
Argon	87.4	83.6	75	38
Nitrogen	77.4	63.4	64	30

*28 mW boils 1 liter of liquid helium per day.

Table 6.3 Typical Thermal Losses for a Two-Stage Cryocooler.

Thermal loss type	First-stage loss	Second-stage loss
Temperature (K)	40	10
Conduction/radiation (W)	17	0.4
Power leads (W)	14	2
Loss margin allowed (W)	7.8	0.6
Total thermal loss (W)	38.8	3.0

mize the temperature differential between the superconducting device and the cryocooler. It is important to mention that the cryocoolers must be designed to handle heat loads with adequate safety margins during the most critical operation. As stated earlier, the transient state operation presents the greatest thermal loads that can limit the duty cycle for a two-stage cooler as shown in Table 6.3. The environmental factors and operating parameters, such as ambient temperature, shock and vibrations, and the availability of proper maintenance service, must be taken into account during the design phase of a cryocooler.

6.4 Microcooler Requirements for Military and Space Applications

Miniaturized cryocoolers, known as microcoolers, have potential applications in space sensors and sophisticated military IR systems, including IR line scanners, IR search and tracking systems, thermal imaging radiometers, and IR missiles. Generally, the microcoolers use the Stirling-cycle engine principle to obtain a liquid nitrogen temperature (77 K) in less than 3 minutes. Such a microcooler incorporates a regenerator to produce a compression-expansion refrigeration system with no valves. The regenerator has a high heat capacity, which acts as a highly efficient heat exchanger. The electronic sensing devices control the cooling action to maintain the required operating temperature of 77 K. A microcooler designed for NASA, which has a thermal-imaging application and uses advanced technologies, weighs only 15 oz, consumes 3 watts of electrical power, has a shelf life of 5 years, and has demonstrated a continuous operation exceeding 8000 hours under space environments. High reliability is possible with the selection of improved materials, self-lubrication features, low-friction clearance seals, the elimination of gaseous contamination, and the use of linear drive mechanism. Split-type Stirling cryocoolers with power ratings from 0.25 to 1 watt using linear drive mechanisms are capable of providing continuous operation exceeding 6500 hours. A microminiaturized cryocooler can provide a cooling capacity of 150 mW at a temperature of 77 K with input power less than 3.5 W, a heat load of 150 mW, and cooldown time ranging from 1.5 min to 120 K to 4.0 min to 77 K from a room temperature of 298 K. Typical power requirements for microcoolers as a function of heat load and operating temperature are shown in Figure 6.7.

6.5 Cryocoolers Using Unique Design Concepts and Materials

This section deals with cryocoolers that use unique design concepts and exotic materials to provide a significant improvement in cooler performance. The design configurations will include improved regenerative heat exchangers and advanced rare-earth materials with significant enhancement in heat capacity.

6.5.1 CRYOCOOLER AND MICROCOOLER DESIGNS INCORPORATING RARE-EARTH MATERIALS

The original two-stage, 4.2-K, G-M/RE cryocooler [4] used rare-earth (RE) materials as regenerative elements, which demonstrated remarkable performance improvement over the existing G-M-cycle coolers with identical design parameters. This particular cryocooler, employing an erbium nickel rare-earth material, can reach a temperature of 4.2 K in a very short time because erbium nickel has a significant heat capacity below 10 K. The normal heat capacity of a G-M/RE cryocooler is about 100 mW at a cryogenic temperature of 4.2 K. However, such a cooler design incorporating a best combination of selected rare-earth materials has demonstrated a heat capacity exceeding 1 W at 4.2 K.

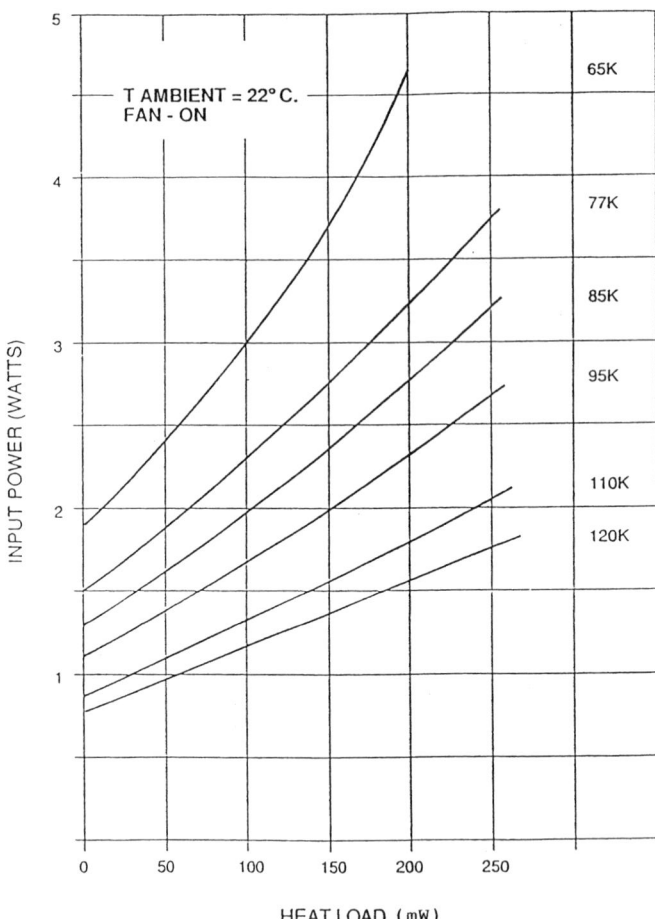

Fig. 6.7 Input power requirements for a cryocooler as a function of heat load and temperature [4].

6.5.2 CRYOCOOLER DESIGN WITH HIGH-PRESSURE RATIO AND COUNTERFLOW HEAT EXCHANGER

Studies performed by the author on cryocooler design configurations indicate that both the high-pressure expansion ratio and the counterflow heat exchanger design are highly desirable to achieve a cryocooler design with low cost, high efficiency, and enhanced reliability. The latest revolutionary Boreas cryocooler design [5] uses both the high-pressure expansion ratio and the counterflow heat exchanger, which eliminates the need for

6. Cryocooler and Microcooler Requirements

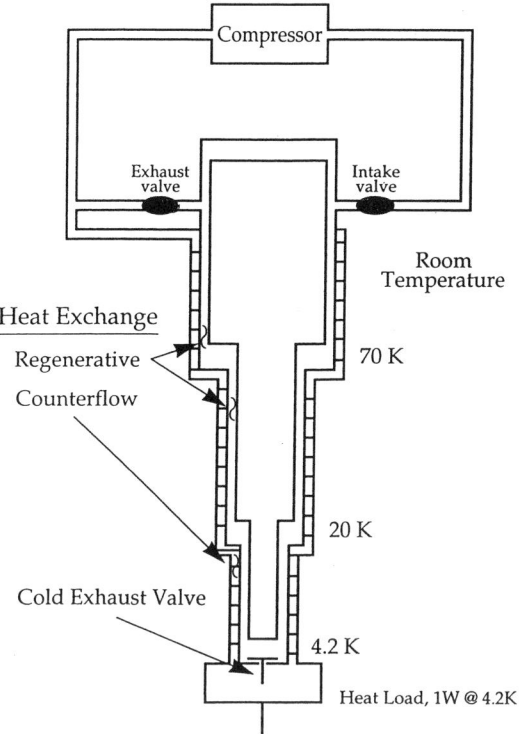

Fig. 6.8 Improved design of a three-stage Boreas cryocooler using advanced technologies.

costly high heat capacity rare-earth regenerator materials to achieve cooling at 4.2 K cryogenic temperature. Note that the Boreas-cycle uses a wet expansion engine to produce liquid helium. The three-stage cryocooler [5] illustrated in Figure 6.8 provides a regenerative heat exchanger function in the warm two stages at 70 K and at 20 K and the counterflow heat exchanger function in the third stage to a reach a low cryogenic temperature of 4.2 K. This particular cryocooler design integrates regenerative and counterflow heat exchanger functions with a three-stage expander to yield a compact cold head with a single helium flow circuit. This unique combination of heat exchanger functions offers a simple, reliable, least expensive cryocooler design with improved performance.

The pressure-volume (P-V) schematic diagram and the 4.2 K-stage efficiency as a function of load temperature are shown in Figure 6.9. It is

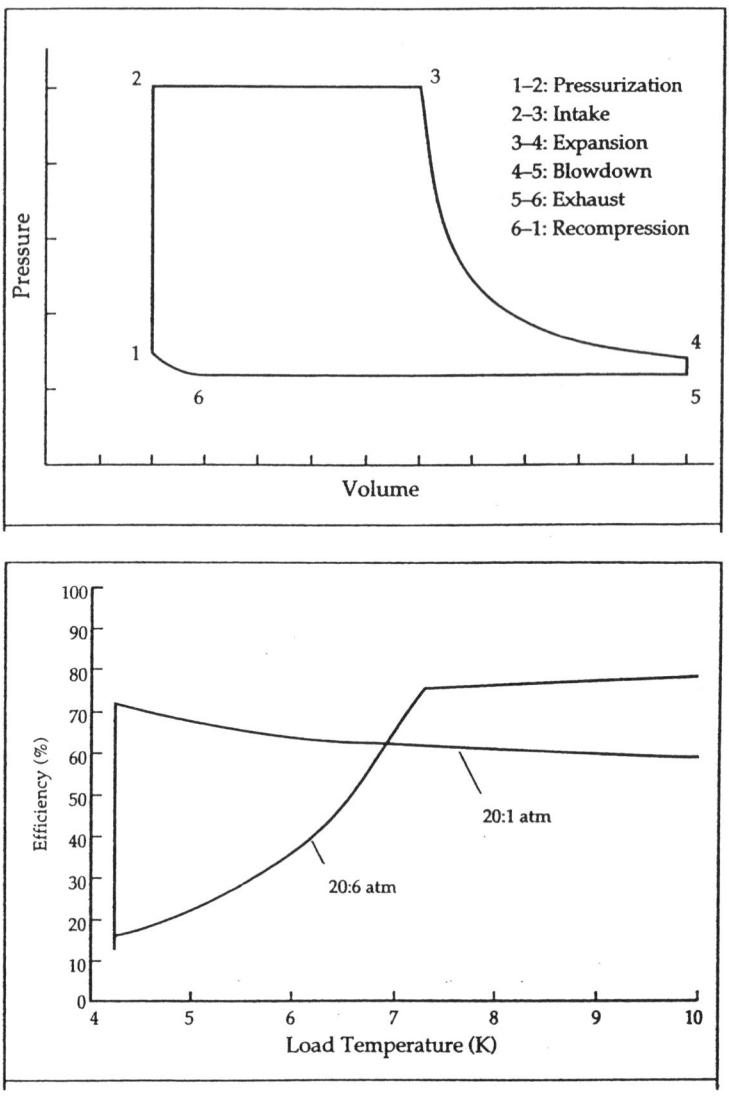

Fig. 6.9 Pressure-volume diagram and cold-stage efficiency of a three-stage Boreas cryocooler [4].

evident from the curves shown in Figure 6.9 that the cold stage efficiency improves with increasing pressure. The expansion cycle reduces the gas temperature. Helium vaporized by the heat loading is returned to the compressor via the counterflow heat exchanger, thereby recuperatively cooling the cylinder and the displacer in preparation for the next operation cycle.

6.6 Critical Thermodynamic Aspects of Cryocoolers

Critical thermodynamic aspects of a cryocooler must be addressed if cooldown time, reliability, and cooling capacity are the critical design requirements. Thermodynamic aspects will be limited to G-M-cycle and Boreas-cycle cryocooler designs to conserve space and time. A rigorous thermodynamic comparison between G-M-cycle and Boreas-cycle cryocoolers both operating at 4.2 K temperature indicates that the Boreas cycle is more than four times more efficient than the G-M cycle. The comparison further indicates that the efficiency of the Boreas cycle is a strong function of the compression suction pressure. The efficiency curve shown in Figure 6.9 shows that the Boreas-cycle efficiency is significantly improved over the G-M cycle at cryogenic temperatures below 7 K operations. This efficiency curve further shows that at 4.2 K operation, the Boreas cycle offers 70% higher efficiency with a suction pressure ratio of 2:1 atm, while the G-M cycle offers an efficiency of 17% at a suction pressure of 20:6. The high efficiency of Boreas cycle is strictly due to the high-pressure expansion ratio to obtain a cryogenic temperature of 4.2 K. The high-pressure ratio is only possible with a counterflow heat exchanger design configuration.

6.6.1 IMPACT OF THERMODYNAMIC EFFICIENCY LIMITS ON VARIOUS COOLING CYCLES

Thermodynamic efficiency is the critical performance indicator of a refrigeration cooling cycle. The thermodynamic efficiency depends on cooling capacity and the input power requirement at a specific cryogenic temperature. The thermodynamic efficiency for a 1-watt, 4.2 K cryocooler can be compared by looking at the power consumption relative to the ideal Carnot cycle. Theoretically, the input electrical power required per watt of refrigeration at a cryogenic temperature of 4.2 K is about 71 watts. A commercially available G-M/J-T cryocooler, which requires 4.5 kW of input power to produce 1 watt of cooling at 4.2 K, offers an ideal efficiency of only 1.6%. It is interesting to note that a G-M/RE (a G-M cryocooler using rare-earth material) cryocooler that uses 0.524 kg of multiple rare-earth materials requires 7 kW of input power to produce 1.05 W of cooling at 4.2 K and operates at 1.1% efficiency of the Carnot cycle. It is now evident which cryocooler configuration offers higher cooling efficiency and consumes higher electrical power for the same amount of cooling under identical cryogenic operating temperatures.

6.7 Techniques to Optimize Cooling Capacity

Several methods or techniques are applied to optimize the cooling capacity of a cryocooler. Maximum cooling capacity is the most important performance parameter of a cryocooler. Discussion on the subject will be limited to a free-displacer, split-type Stirling refrigeration system (Figure 6.10) because of its several outstanding advantages. The maximum available cooling (MAC) capacity of a cryocooler can be optimized for a specific application and under specific operating environments. Critical elements of a split-type Stirling cryocooler are specified in Figure 6.10. The MAC capacity for this particular cryocooler can be calculated by integrating the expansion space represented by the pressure-volume curve. In other words, the MAC capacity is proportional to the integral of the expansion pressure and the displacer volume. However, the net cooling capacity is equal to maximum cooling capacity less the heat conduction loss of the regenerator, the enthalpy flow loss of regenerator, the shuttle loss of the displacer, and the hysteresis loss of the gas. Cryocooler design analysis [5] using linear network modeling indicates that the net cooling ca-

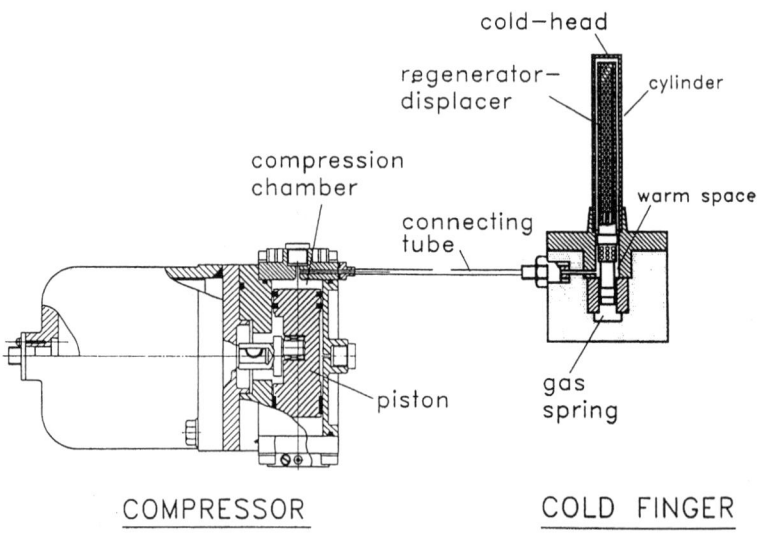

Fig. 6.10 Critical elements of a Stirling refrigerator design with optimum cooling capacity [5].

pacity (Q_{net}) increases with the increasing cold-end temperature and with a decreasing displacer loss coefficient (C_d). The linear model further indicates that for a given displacer loss coefficient, the net cooling capacity first increases with increasing frequency reaching a peak and then decreases as illustrated in Figure 6.11. An optimum value of net cooling capacity occurs [5] when the normalized natural frequency has a value between 1.2 to 1.5, depending on the value of the displacer loss coefficient.

The loss coefficient depends on the displacer seal design, the seal material used, and the quality of the workmanship. Accurate performance prediction is possible through a modified linear network analysis [5]. The loss coefficient must also take into account the frictional and gas leakage losses of the displacer. This means that the loss coefficient is not constant, but it can vary as a function of displacer oscillating frequency (f) and cold-heat temperature (T_L). The net cooling capacity as a function of the correlated value of the loss coefficient is illustrated in Figure 6.11. Note that the upper curve shown yields a more accurate estimate for the net cooling capacity of a cryocooler as a function of cold-end temperature.

6.8 Optimization of Temperature Stability and Mass Flow Rate

The transient behavior of a cryocooler is of critical importance, particularly, when the cooler is integrated and operated in a complex space or military system. Predicting the transient behavior of a self-regulating cryocooler can be accurately accomplished through a numerical simulation. In other words, transient behavior can be predicted by modeling various types of bellow control mechanisms deployed in a self-regulating cryocooler design configuration. Optimization of temperature stability depends on the thermal performance of critical elements used in the design of a control mechanism.

Note that a flow regulating mechanism in a cryocooler offers long-term, low gas consumption with maximum economy. A bellow control mechanism is widely used for the gas flow regulation. The temperature-sensitive bellow mechanism senses the temperature at the cold end and then regulates the opening of the cooler orifice, thereby allowing the necessary gas flow to maintain a specified cold temperature and to minimize the excess gas flow to avoid wasting the refrigerant. Transient simulation

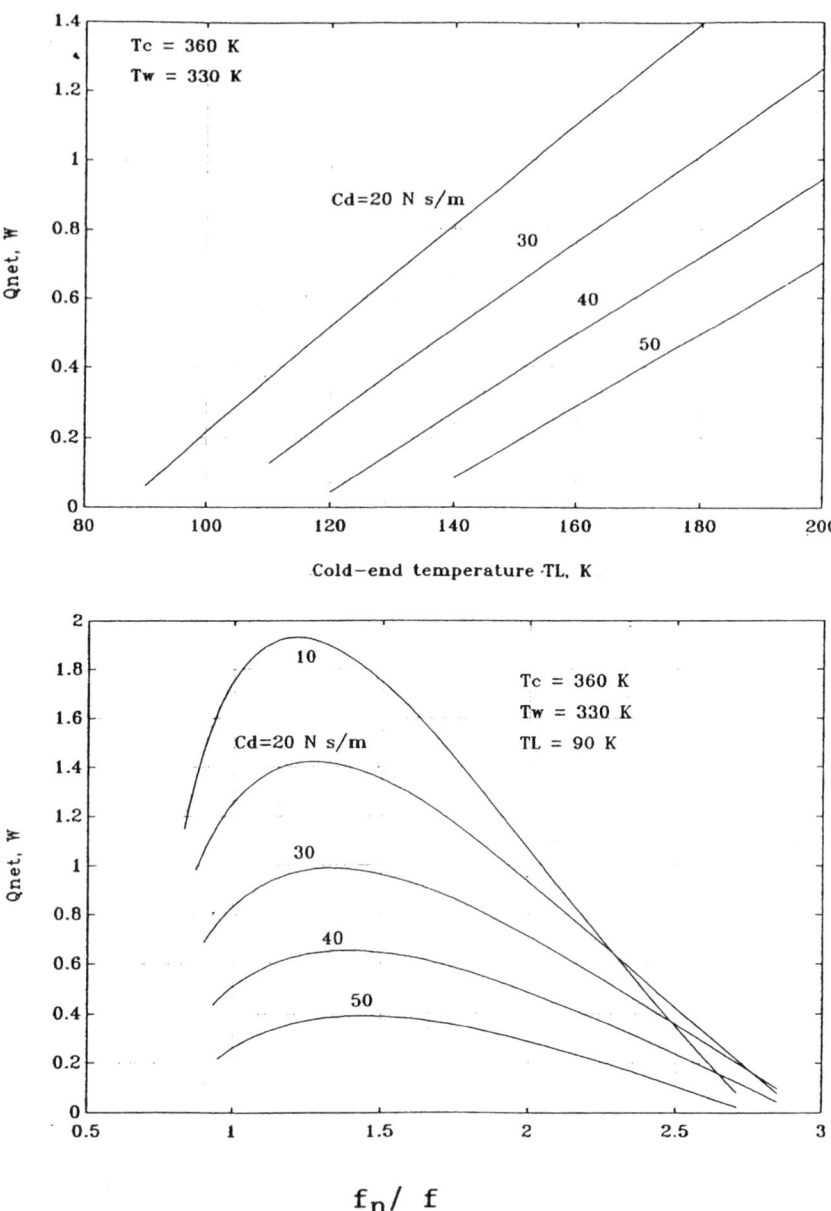

Fig. 6.11 Net cooling capacity variation as a function of cold-end temperature, frequency and displacer loss coefficient, C_d [5].

requires some critical parameters such as operating temperature, mass flow rate, orifice opening, and the nose area ratio (NAR) parameter of the orifice. The mass flow rate of the gas is measured in kg per second. The temperature-sensitive bellow mechanism allows a high flow rate to meet fast cooling requirements and then throttles down to conserve the cooling gas. The design configuration of the needle valve provides control or variation of the valve opening, which regulates the gas flow rate. The NAR parameter as a function of the mean operating temperature of the gas inside the bellows indicates the level of valve opening. The parameter NAR is defined as the area ratio of maximum valve opening to a specific valve opening to meet a specific gas flow rate. The spatial temperature variations for high-pressure gas and the glass Dewar assembly depend on assembly position with respect to the initial position. The time required to drop a temperature to a specified value is a function of the valve opening, the type of bellow mechanism deployed, and the type of gas used by the cooler. Typically, it will take about 30 seconds for the temperature to drop from 300 K to 80 K. It will take longer for the temperature to drop from 300 K to 45 K or lower.

6.9 Cryocooler Design Requirements for Space Applications

Published literature on space-based cryocooler design configurations indicate that radiant coolers are best suited to maintain required operating temperatures for various electronic, infrared, and electrooptic devices and sensors operating in a spacecraft. These radiant coolers are designed to maintain operating temperatures in the 80 to 200 K range depending on the physical dimensions of the reflector surfaces. In general, the low-temperature radiant coolers deploy a conical or rectangular second-stage structure with high reflectivity. This high-reflection surface provides adequate thermal shielding for the cold patch from space and earth and reflects solar radiation at a shallow angle reaching the patch. A two-stage radiant cooler with a rectangular configuration is best suited for a low-earth orbiting (LEO) satellite or spacecraft.

Sensors operating in deep space can expect operating temperatures ranging from 80 to 220 K. The low effective sink temperature of deep space offers the most ideal natural environment for the passive radiant cooling of IR detectors and devices. Note that radiant cooling involves no moving parts, offers high mechanical integrity in space environments, and

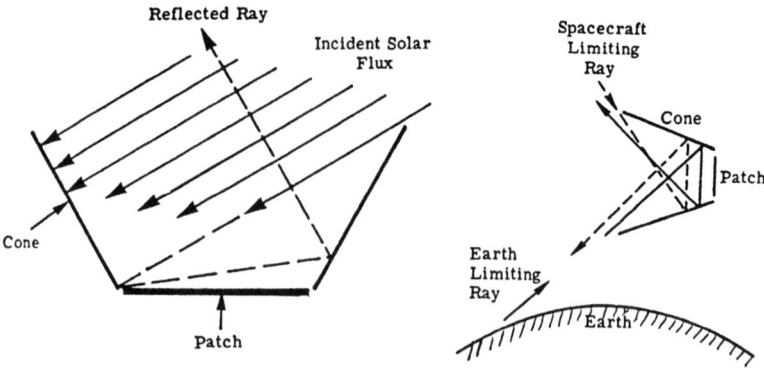

(A) Critical elements of a radiant cooling system

(B) Reflection from earth and spacecraft

Fig. 6.12 Incident solar flux and reflected rays associated with a radiant cooling system attached to a spacecraft.

requires no electrical power. The effective deep-space temperature is around 4 K, and thus one or more IR detectors mounted on a cold plate with appropriate surface and high emissivity can radiate efficiently to the deep-space environment at a temperature of 4 K. The presence of high vacuum at orbital altitude minimizes the efficiency of convectional heating in space. The high efficiency of a radiant cooling system requires the cold plate to be shielded with a conical structure against the heat energy from thermal emission and sunlight reflections from the earth in case of near-earth orbit operation. The cold plate shown in Figure 6.12 must be thermally shielded from the space environments to achieve high cooling efficiency.

The radiant cooling system design must take into account several operational parameters including type of satellite (near-polar, equatorial, or elliptical), orbiting altitude, the orientation of the spacecraft or satellite relative to earth, and the location of the radiator on the spacecraft. It is of paramount importance that the radiant cooling patch surface for detector mounting is large enough so that the thermal energy produced by the attachment of IR detecting elements and associated electronic circuits is small enough compared to total power radiated by the patch surface at its equilibrium temperature. Note that thermal-balance equations (TBEs) for both the patch and the cone shown in Figure 6.12 are needed to determine

Table 6.4 Typical Performance Characteristics of Space-Based Radiant Coolers.

Space system number	Orbit	Altitude (nm)	Temperature (K)	Net cooling capacity (mW)	Patch area (cm^2)	Weight (lb)
1	NPSS	600	195	1.7	9.8	1
2	NPSS	500	110	4.7	23	9
3	NPSS	495	95	1.0	46	16
4	NPSS	625	120	8.0	23	9

Note: Symbol NPSS stands for a spacecraft launched in the near-polar sun-synchronous orbit.

the equilibrium temperature of the patch. The TBE for a black, isothermal patch of an ideal radiant cooler mounted on a spacecraft is dependent on the conductive cooling efficiency and IR detector-thermal inputs to the radiator input terminals. The TBE assumes no direct environmental-radiative inputs to the patch, but only to the cone and its supporting structure. The TBE for the cone can be determined by equating the radiative heat output of the cone to both the radiative and conductive heat input. It is important to point out that the heat transferred from the cone surface to the patch surface is small compared to heat energy emitted and absorbed by the cone surface. The second-stage reflector with high reflectivity not only shields the cold patch from the space radiation and interaction between the earth and space but also reflects solar energy at shallow angles away from the cooler before it reaches the patch. In most cases, well-designed sun and earth shields are provided to minimize parasitic heat loading of the patch. Note radiators with large patch surface areas provide cooling capacities between 1 and 5 watts and can have a major impact on the spacecraft design configuration. However, the net cooling capacities for space-based coolers with small surface areas hardly exceed 10 mW, as is evident from the data summarized in Table 6.4 [6].

6.10 Summary

Cryogenic systems or devices cannot operate efficiently without integrating cryocoolers with appropriate cooling capacities. A cryogenic cooling subsystem is an important and costly component of a superconducting system. The weight, size, cost, and complexity of a cryocooler depend on the operating cryogenic temperature, the minimum time to achieve the desired cryogenic temperature, and the cooling capacity to meet the system performance requirements. This chapter summarized the performance capabilities and limitations of various cryocoolers and microcoolers used in space, medical, military, and research applications. Critical operating environments that impact the maintenance and reliability aspects of a cryocooler were discussed. Regular maintenance and servicing requirements necessary to ensure safe, steady, and reliable operation of a cryocooler were defined. The overall cryogenic system cost depends on several factors including the reliability, field maintenance, frequency of refrigerant refills, and operating environments. Input power requirements of a cryocooler as a function of cooling capacity and cryogenic temperature were specified for various applications. Critical design aspects for microcoolers for airborne, space, and missile applications were summarized. In the case of space applications, cryocooler or microcooler weight, size, and reliability over extended periods are of paramount importance. Performance requirements of cryocoolers for IR detectors, focal planar arrays, photonic devices, and electrooptic sensors were briefly summarized, with an emphasis on cooling capacity and reliability. The performance characteristics of space-based radiant coolers were described, with a particular emphasis on cooling capacity, reflector configuration, and patch size.

References

1. A. P. Auckermann. "Closed-cycle refrigerators for superconducting applications." *Superconductor Industry*, Fall 1993, pp. 14–44.
2. A. R. Jha. *Superconductor technology: Applications to microwave, electrooptics, electrical machines and propulsion systems*. New York: John Wiley & Sons, 1998, p. 281.
3. Jha, *Superconductor technology*, p. 283.
4. M. Wilson. "Advances in 4.2 K crycooler technology." *Superconductor Industry*, Fall 1993, pp. 30–35.
5. Jha, *Superconductor technology*, p. 290.
6. W. L. Wolfe and G. J. Zissis (Eds.) *The Infrared Handbook*, 1978, radiant-cooling pp. 15–46.

Chapter 7 | Integration of Latest Cooler Technologies to Improve Efficiency and Cooling Capacity

7.0 Introduction

This chapter focuses on the integration of the latest cryocooler technology capable of improving efficiency, enhancing reliability, and providing high cooling capacity with minimum cost and complexity. Unique design concepts will be integrated in the cryogenic coolers for industrial systems, complex military sensors, and medical diagnostic equipment. Design concepts capable of yielding improved performance over extended durations will be discussed, with a major emphasis on cost and reliability. Pulse tube refrigerator (PTR) design aspects are examined in greater detail because of their potential applications in missile, airborne, and space systems, where compact size, light weight, and minimal power consumption are critical. The latest research and development activities indicate that PTR technology offers a cryocooler with minimum cost, size, and weight. According to some scientific publications, the PTR cooling scheme involves the latest refrigeration technology with only prevalent Gifford-McMahon (G-M) and Stirling cycle refrigerators. Performance requirements for high-power cooling systems in sonar transmitters, linear accelerators, medical diagnostic equipment such as magnetic resonance imaging and computer tomography, and megawatt-class lasers such as the chemical oxygen iodine laser (COIL) operating at 1.315 micron are briefly summarized.

7.1 Unique Design Concepts and Advanced Materials

PTR technology has been recognized as the hottest and the latest cooler technology best suited for G-M and Stirling-cycle refrigeration systems. The PTR cooler represents the advanced cooler design with no moving parts in the cold head and offers reliable and vibration-free operation that is most suited for highly sensitive applications including airborne and space IR sensors, satellite communications and wireless systems, superconducting quantum interference device (SQUID) sensors, superconducting magnet coils for MRI equipment, scientific research instruments, and cryopumps for high density integrated circuit (IC) fabrication. Advanced regenerative rare-earth and high-performance plastic materials with outstanding thermal and mechanical properties are identified that will significantly improve the overall performance level of a cryocooler. Integration of multistage-buffer technology, low-temperature switching techniques, and dual-orifice design concepts are discussed in great detail. The implementation of unique design concepts, latest techniques, and advanced materials will significantly improve the adiabatic efficiency, reliability, cooldown time, and cooling power of a cryocooler.

7.2 Design Concepts for a Pulse Tube Refrigerator (PTR) Cryocooler

Significant research and development activities currently focus on various means and ways that will significantly improve the performance of a PTR cryocooler with minimal cost. Studies performed by the author indicate that a PTR system is best suited for applications, where high reliability, improved adiabatic efficiency, and vibration-free operation are the principal requirements. Important components of a PTR cryocooler are illustrated in Figure 7.1. It is important to mention that the oscillating pressures inside a pulse tube closed at one end will cause the gas to be heated at the closed end during the compression cycle and cooled at the other end during the expansion cycle. The expansion and compression of the refrigerant or gas provide the cooling effect, as it is with other refrigeration systems. Note that the heat exchange between the gas and the pulse tube walls is primarily responsible for improving the cooler efficiency [1]. The absence of moving parts in the cold head in a PTR system offers high reliability, low maintenance cost, improved adiabatic efficiency, minimal noise, and vibration-free operation not matched by any other cryogenic cooler to date.

Implementation of an orifice inside the warm end of the pulse tube causes the gas flow in phase with the operating pressure and brings the temperature down to 105 K. This results in an increase in refrigerator

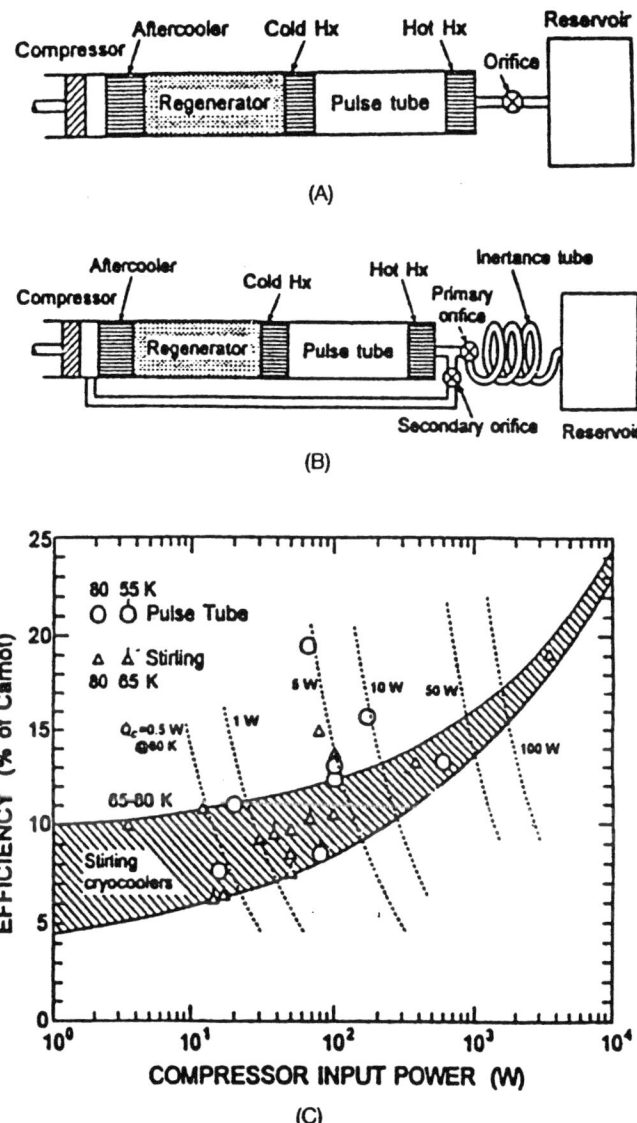

Fig. 7.1 Block diagram of pulse tube refrigerator (PTR) with (A) single orifice, (B) two orifices and inertance tube, and (C) efficiency versus compressor input power requirements for cryocoolers [1].

power. However, introduction of a second orifice (Fig. 7.1B) in the "double-inlet" section brings the cooler efficiency equal to that of a Stirling-cycle refrigerator of comparable size, which indicates a major breakthrough for commercial applications. Note that a PTR design with the double-inlet feature can produce a heat lift or cooling power of 30 watts at 80 K operation with an efficiency of 13% of the Carnot cycle (the Carnot cycle represents the efficiency of an ideal refrigeration system). Even the smallest PTR system with a heat load of 0.5 watt demonstrated efficiency better than 8%. An improved version of a PTR system incorporating a critical component called the "inertance tube" (Fig. 7.1C) demonstrated an efficiency close to 19% of the Carnot cycle, which is significantly higher than the best Stirling-cycle refrigerator shown in Figure 7.1(C). Efficiency and compressor input power for Stirling-cycle and PTR coolers are compared in Figure 7.1(C).

7.2.1 PERFORMANCE CAPABILITIES OF A PTR SYSTEM

The performance level of a PTR system can be improved by selecting an optimized design configuration with proper tube orientation with respect to gravity and by using helium gas as a working fluid at optimum pressure. A PTR system with 6-watt rating at 60 K and with a 1000-watt compressor unit has demonstrated remarkable performance improvement by selecting optimum design configuration. A thermodynamic model developed for a single stage PTR system indicates an optimum performance for a given heat lift at operating temperatures exceeding 30 K. The thermodynamic model further reveals that the heat transfer between the gas and the tube walls within the pulse tube determines the lowest physical size of the pulse tube. The model further predicts that a PTR system using optimum design parameters can achieve a cooling power as high as 30 watts at 60 K with a compressor rating not exceeding 1000 watts. Lower temperature operations are possible with a multistage PTR system. For example, a three-stage PTR cooling system designed by a Japanese professor, Matsubara, in 1994 demonstrated an operating temperature as low as 4 K. Furthermore, in 1997 a German scientist developed a two-stage PTR system setting the world's record for the PTR performance level at 2.2 K. A minipulse tube recently developed by TRW demonstrated a cooling power of 0.8 watt at 80 K with minimum weight and size. This kind of mini-PTR cooling device is best suited for space lab and reconnaissance satellite applications where weight and size are critical.

Lockheed-Martin Astronautics Division [1] developed a miniature PTR cooling system with a cooling power of 50 mw, swept volume of 0.75 cu·cm³, and mechanical input power less than 10 watts. This smallest PTR cooler is best suited for cooling IR sensors or devices deployed in complex missile systems, satellite communication equipment, and space reconnaissance sensors, where electrical power consumption, weight, and size are of paramount importance.

7.2.2 THERMODYNAMIC ASPECTS OF A PULSE TUBE CRYOCOOLER (PTC)

The optimum PTC design requires complete familiarization with the basic thermodynamic aspects and thermal equations associated with a specific design configuration of the PTR system. Furthermore, thermal aspects of Stirling-cycle and G-M-cycle cooling systems need to be reviewed with major emphasis on the reversible and irreversible effects in the regenerator assemblies of these coolers. The thermal treatment on the pulse tube involves derivation of energy balance equations for the gas used as a working fluid, the matrix in the regenerator, and other critical properties of the components deployed by the PTR system shown in Figure 7.2.

7.2.2.1 Derivation of Expressions for Various Operating Parameters

The ideal heat exchange between the working fluid or gas and the matrix of the regenerator has a major impact on various equations that have solutions in which the temperature profile moves through the regenerator as a wave. Entropy production expressions for the regenerator orifices and heat exchanger need to be developed. A schematic diagram of two distinct pulse tube configurations — Stirling- and G-M-type coolers — are shown in Figure 7.2 with the relevant thermodynamic quantities.

For a gas with molar internal energy U_m, molar volume V_m, molar entropy S_s, molar enthalpy H_m, and molar Gibbs free energy parameter μ_G at a given pressure P and temperature T, one can write the expressions for preceding parameters for the gas as

$$\text{For a molar enthalpy, } H_m = [U_m + PV_m] \quad (7.1)$$

$$\text{For Gibbs free energy parameter, } \mu_G = [H_m - TS_m] \quad (7.2)$$

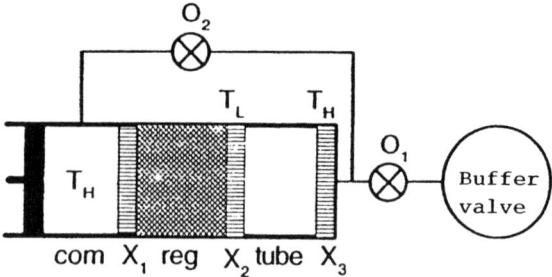

(A) Pulse tube refrigerator design using the double-orifice technique (based on Stirling-type cycle)

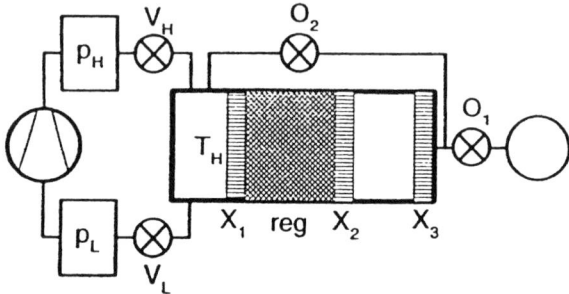

(B) Schematic diagram of a pulse tube refrigerator (PTR) using the G-M cycle with the double-orifice design configuration

Fig. 7.2 Schematic diagrams of PTR system using the various-cycle operating principle and double-orifice configuration. Symbols: T_h = high or room temperature (300 K); T_c = low temperature (K); X_1 = room temperature heat exchanger; X_3 = last room temperature heat exchanger; X_2 = low temperature heat exchanger; O_1, O_2 = orifice no. 1 and no. 2; and V_H, V_L = high pressure and low pressure valves.

$$[dS_m] = [(C_p/T)dT] - [(\partial V_m/\partial T)_p dP] \qquad (7.3)$$

$$[TdS_m] = [dH_m - V_m dP] \qquad (7.4)$$

$$[dH_m] = [C_p dT] - [V_m - T(\partial V_m/\partial T)_p dP] \qquad (7.5)$$

$$[dU_g] = [T_g dS_g - PdV + \mu_G dN] \qquad (7.6)$$

where, N is the number of moles in the regenerator section, T_g is the gas temperature, and S_g is the entropy of the gas at its operating temperature.

These equations yield the values of internal energy (U), enthalpy (H), and entropy (S) in the various regions of the cryogenic refrigerator. Studies performed by the author on the subject reveal that the rate of increase in the internal energy of the gas in a thermodynamic system with volume V is equal to the algebraic sum of the heat flow (Q) and the enthalpy flows (H) minus the work done. Note that the work done needed to remove the heat amount Q_L at a constant low temperature T_L during one cycle (Figure 7.3) and to release the heat at high or ambient temperature T_H can be written as

$$W = [(T_H/T_L - 1)Q_L + T_H S_i] \qquad (7.7)$$

where, S_i indicates the sum of the entropy contributions produced by all irreversible processes in the cryogenic cooler during one operating cycle. The total amount of work done by the compressor unit (Wc) per cycle can be written as

$$Wc = [\int A_p P_p v_p dt], \text{ when integrated over the time period of the cycle } (t_c) \qquad (7.8)$$

These expressions can be used as guidelines to obtain a practical design of a regenerator, but not for designing the entire cooler. Furthermore, the expression for the entropy provides a realistic estimate for the entropy produced in various regions of the pulse tube. The entropy produced due to heat transfer during one cycle is due to contributions from four distinct sources; the heat transfer between the gas and the matrix, the thermal conductivity of the gas, the thermal conductivity of the matrix (or regenerative materials in the regenerator), and the flow resistance in the matrix. During the compression and expansion cycles, various types of losses occur including orifice losses, fluid losses, valve losses, heat exchanger losses, wear and tear losses in the drive mechanism, and regenerator losses. The heat exchanger losses are strictly due to heat transfer over a finite temperature difference. Under turbulent gas flow conditions, entropy is produced when two gas elements containing N_1 and N_2 moles of gas at temperatures T_1 and T_2, respectively, mix adiabatically to form uniform gas at a constant temperature and a constant pressure. The entropy production in a regenerator is due to a finite heat exchange between the gas and the matrix.

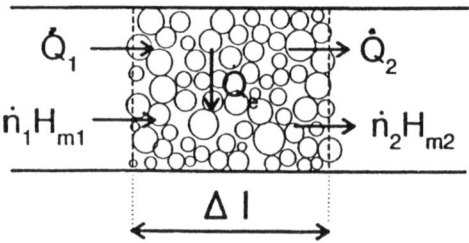

(A) Heat energy flow (Q) in the regenerator section of thickness delta-l

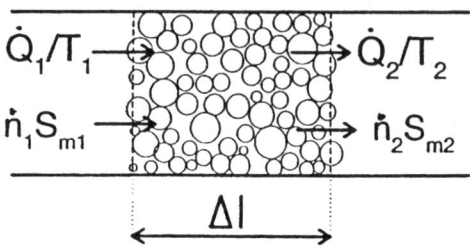

(B) Entropy flow (S) in the regenerator section of a PTR

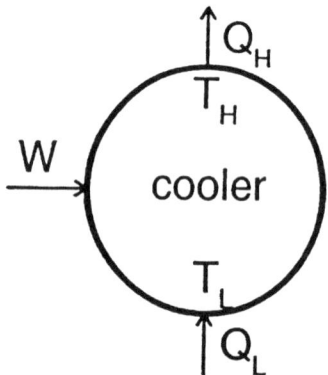

(C)   Parameters during work done in one cycle

Fig. 7.3 Schematic diagrams showing (A) heat flow, (B) entropy flow, and (C) work done in one cycle of a PTR system. Symbols: Q_L = heat flux at low temperature, T_i; Q_H = heat flux at high temperature, T_H; H_{m1}, H_{m2} = molar enthalpy for orifices 1 and 2, respectively; S_{m1}, S_{m2} = molar entropy for orifices 1 and 2, respectively; and $\dot{n}_1$, $\dot{n}_2$ = molar entropy for orifices 1 and 2, respectively.

7.2.3 MINIMUM REFRIGERATION TEMPERATURE (MRT) OF A CRYOCOOLER

Minimum refrigeration temperatures (MRTs) are possible with liquid helium cooling agents. In addition, in traditional regenerators much lower MRTs are possible with helium II (He II) and using advanced rare-earth materials in the regenerator. Studies performed by the author indicate that the isentropic expansion at constant entropy and the isobaric specific heat of the refrigerant at constant pressure determine the upper limit of the MRT. The refrigeration temperature, to a great extent, depends on the adiabatic expansion on the working fluid in the dedicated expansion space. The isentropic expansion indicates the temperature change due to a pressure change at constant entropy.

The studies further indicate that the lowest MRT can also be achieved through the isentropic compression process instead of through isentropic expansion as illustrated in Figure 7.4. Various thermodynamic processes such as isentropic expansion, isobaric heat absorption, isentropic compression, and isobaric heat rejection associated with He II cryocoolers [2] are shown in Figure 7.4.

7.3 Ways and Means to Improve PTR Performance

This section deals with various techniques capable of achieving significant improvement in the performance level of a PTR system. As stated earlier, the PTR cryocooler design promises significant advantages over the conventional cryocooler in terms of reliability, cost, efficiency, longevity, and vibration-free operation. Deployment of the latest magnetic rare-earth materials such as Er_3Ni has provided large volumetric specific heat near 4 K operation, leading to significant increase in cooling power and regenerator efficiency as illustrated in Figure 7.5. A lowest operating temperature of 2.3 K has been demonstrated by a PTR cryocooler by incorporating a second stage regenerative tube filled with three distinct materials: ErNiCo, ErNi, and Pb. This particular PTR system provided a cooling power of 370 mw at 4.2 K and 700 mw at 5 K operations with an operating frequency of 1 Hz and helium pressure of 1.55 Mpa or 224.75 psi. The cold head of this particular system reached the cryogenic temperature of 4.2 K within 100 minutes. This type of pulse tube is most suitable for low temperature superconducting devices operating below 4 K,

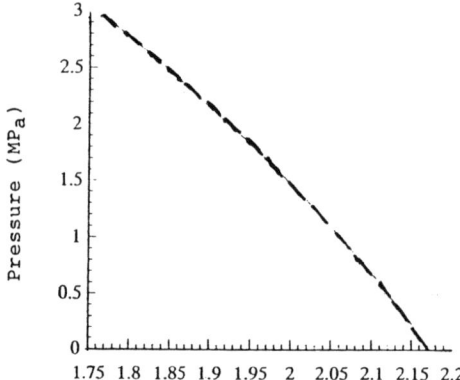

(A) Minimum refrigeration temperature as a function of pressure definition of various processes:

process 1-2 = isentropic compression; process 2-3 = isobaric heat absorbing; process 3-4 = isentropic expansion; and process 4-1= isobaric heat rejection

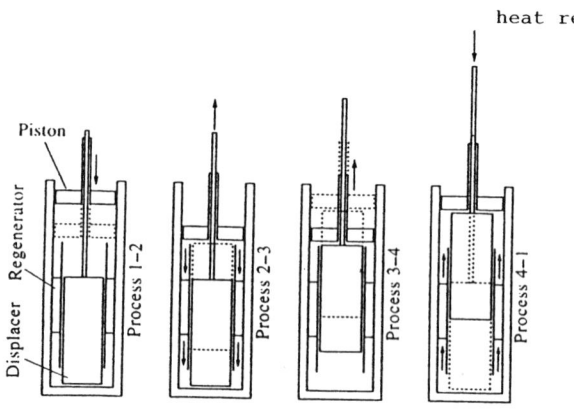

(B) Schematic of a Helium II cryocooler showing various processes involved

Fig. 7.4 Minimum refrigeration temperature and description of various processes involved in the helium II cryocooler.

because its second-stage temperature reaches well below 4 K in the shortest time with a cooling power capability in excess of 200 mw at a pressure of 225 psi and at an operating frequency of 1 Hz as illustrated in Figure 7.6. The same tube offers a cooling power of 2 W at 45 K. Cooling capacity as a function of first- and second-stage operating temperature is evident from Figure 7.6.

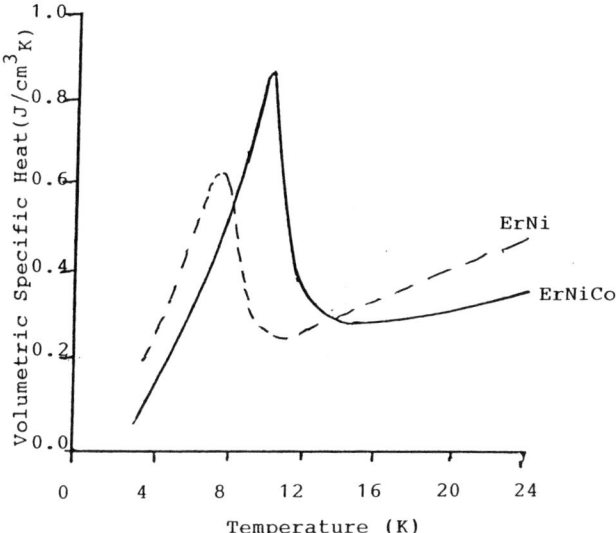

(A) Volumetric specific heat of regenerative materials in second stage

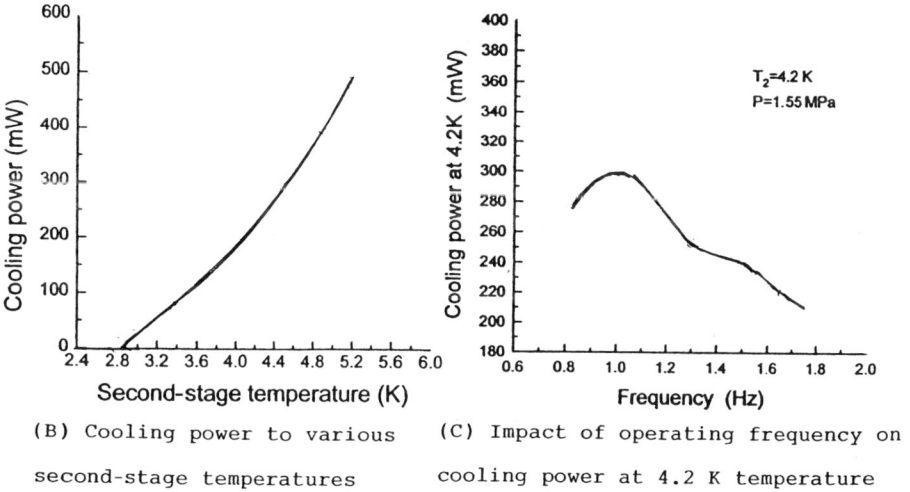

(B) Cooling power to various second-stage temperatures

(C) Impact of operating frequency on cooling power at 4.2 K temperature

Fig. 7.5 Performance of second stage with two-layer regenerator and with optimum settings of orifice and double-inlet values when the regenerator is filled with regenerative materials.

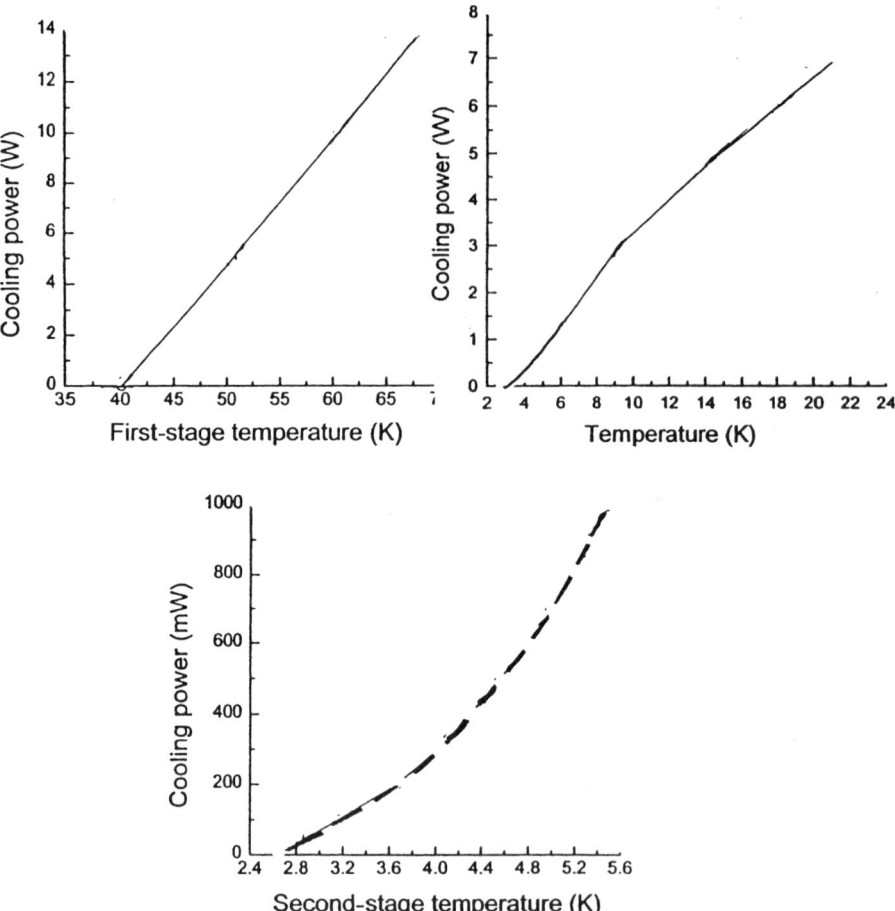

Fig. 7.6 Cooling power for the first stage and second stage as a function of their respective operating temperatures when the valve settings are optimized for maximum cooling power at 4.2 K.

7.3.1 IMPACT OF COOLANT AND REGENERATIVE MATERIALS ON COOLER PERFORMANCE

PTR system performance is dependent on the type of working fluid used and the regenerative materials selected in the regenerator. Studies performed by the author reveal that the second-stage design must take into account the properties of the helium as a working fluid. Furthermore, the optimum length of the second stage regenerator tube and type and amount of regenerative rare-earth materials must be selected for a given volume. Selection of the coolant, type and amount of regenerative materials, and axial length of the regenerator has a major impact on the refrigeration efficiency and cooling capacity of a PTR system. The studies further reveal that there is axial heat conduction along the pulse tube walls that is in close contact with regenerative materials. A regenerative tube between the warm end of the pulse tube and the room temperature must be used to avoid the high-temperature gradient and to improve the performance of the coldest section of the tube. Estimated cooling power as a function of operating frequency and second-stage temperature is shown in Figure 7.5, when the second-stage tube is filled with high-performance regenerative materials, namely, ErNiCo, ErNi, and Pb. Note the cooling power capacities of the first- and second-stage cold heads can be optimized at a specified operating temperature by selecting the right operating frequency and gas pressure. For example, the cooling power of the second stage can be as high as 325 mw at 4.2 K with the first stage operating at a temperature of 52 K with a heat load of 5 W. The increase in cooling power at 4.2 K operation can be achieved by increasing the first-stage heat load. Such a procedure will cause a larger mass flow rate through the second-stage cold head at a higher temperature than the first stage. Higher cooling capacity is also possible through other means such as proper timing of the rotary valve and optimization of the coldest regenerator section.

7.3.2 PARAMETRIC ANALYSIS TO PREDICT PTR SYSTEM PERFORMANCE

Studies performed by the author indicate that the performance of a PTR system can be optimized through parametric analysis using computer simulation techniques. Due to the complex design of a PTR cooler, computer simulation will be most desirable in predicting and optimizing the overall performance level as a function of various operating parameters and physical dimensions of the regenerator. As stated earlier, the PTR system involves a buffer section with specified volume, orifice valves, cold heat, a regenerator section filled with exotic rare earth materials, and a phase shifting element to introduce a specified phase shift between the gas pressure. Note that the minimum refrigeration temperature of a cooler in a given time is dependent on the number of stages.

Computer simulation predicts that a PTR cooling system with a single stage can provide an operating temperature as low as 24 K without involving regenerative materials. However, introduction of advanced regenerative materials in the coldest part of the cooler can provide an operating temperature of 10 K. An operating temperature as low as 4 K can also be achieved through a multistage PTR design, but at higher cost and complexity. A three-stage PTR system designed by a Japanese scientist, Matsubara, reached a cryogenic temperature of 3.6 K with a net cooling power of 30 mw at 4.2 K. The latest PTR design survey indicates that a two-stage PTR cooler reached the lowest temperature of 2.3 K with a cooling power of 370 mw.

7.3.3 IMPACT OF REGENERATIVE MATERIALS ON COOLER PERFORMANCE

Potential regenerative materials include an erbium-doped nickel-cobalt compound (ErNiCo), erbium-doped nickel (ErNi), and lead metal (Pb). The first two materials offer higher performance levels. The regenerative materials with proper geometrical shapes are located in the regenerator at specified locations to meet the specific operating temperature and desired efficiency requirements. Physical dimensions of a regenerator also play a key role in meeting specific cooler performance requirements. A two-stage pulse tube with an inner diameter of 15 mm, a length of 350 mm, and a reservoir volume of $350\,cm^3$ incorporating a regenerator with an inner diameter of 29 mm and length of 180 mm filled with ErNi rare-earth material can offer a first-stage temperature of less than 56 K and a second-stage temperature of 4 K with an impressive cooldown period. However, these operating temperatures are possible at an operating frequency of 1.66 Hz and a gas pressure of 261 psi. The introduction of a phase shifter in the second stage of the pulse tube offers increased regenerator performance by decreasing the pulse tube losses. Note that a double-inlet phase shifter, when incorporated in the double-orifice design configuration, provides improved phase shift effect. In brief, deployment of the latest regenerative materials, optimization of the pulse tube size at each stage, and selection of right operating parameters, such as mean pressure, pressure ratio, and operating frequency, can significantly improve the overall PTR system performance.

It is important to mention that the temperature distribution of the regenerative material varies from time to time during one cycle period of the pressure waveform. For the ErNi material, a minimum refrigeration temperature of 4.2 K occurs when the rare-earth material is located at a distance about 70% away from the input end of the regenerator as illustrated in Figure 7.2. Note that the increase in the volumetric specific heat of helium occurs in the middle of the regenerator section of the pulse tube. The efficiency of the regenerator is strictly dependent on the ratio of heat capacity of the helium working fluid to the heat capacity of the matrix or regenerative material used. In summary, the performance of a low-temperature pulse tube cooler depends on the properties of the regenerative materials and the heat transfer in the heat exchanger and regenerator sections illustrated in Figure 7.2.

7.3.3.1 Impact of Phase Shift on Cooling Performance

As stated earlier, the phase shift between the working fluid pressure and velocity is very vital to maintain high cooler performance in a pulse tube refrigerator. This performance depends on the gas-to-liquid ratio, which varies from 670:1 for air to 650:1 for nitrogen to 700:1 for helium. The higher this ratio, the better the cooler performance in general. A typical phase diagram showing the gas, liquid, and solid phase regions for nitrogen as a function of operating temperature and pressure is depicted in Figure 7.7.

7.3.3.2 Impact of Heat Leakage on Cooler Efficiency and Cooldown Time

Heat leakage from the Dewar assembly and cooler housing can decrease the cooling efficiency of a cryocooler, regardless of cold-end temperature. Cooler capacity and Dewar heat leakage as a function of ambient temperature are shown in Figure 7.8. It is evident from this figure that Dewar heat losses increase with the increase in ambient temperature. Figure 7.9 shows that the second-stage load increases with the increase in cold-end temperature. The cooldown time for both the first and second stage increases with the decrease in the operating temperature. This means that larger cooldown times are required to achieve lower cryogenic temperatures. Typical cooldown time for a second-stage, 0.25-W split Stirling-cycle cryocooler from room temperature (300 K) to 25 K operating temperature is about 6 minutes, as illustrated in Figure 7.9. It is important to mention that cooldown time will increase with higher heat losses regardless of the leakage sources.

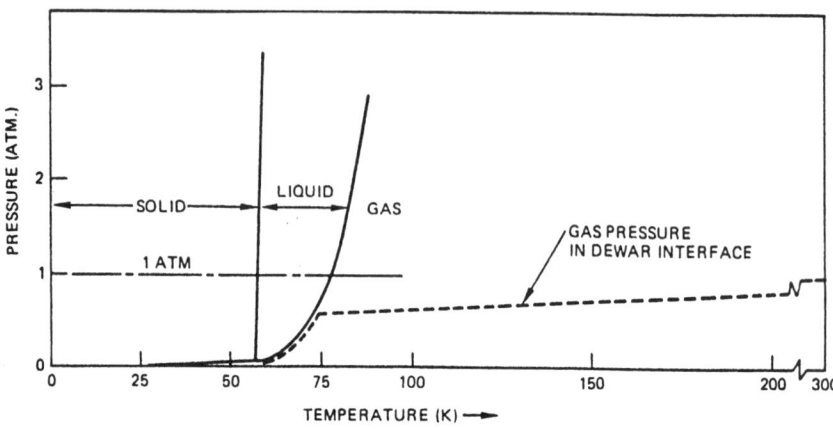

PHASE DIAGRAM FOR NITROGEN

Fig. 7.7 Typical phase diagram for nitrogen showing the gaseous, liquid, and solid phase regions as a function of cryogenic temperature.

7.4 Cryocooler Designs for Industrial Applications

Cryocoolers for industrial applications are designed with large cooling capacities, which require relatively higher operating temperatures. The latest research and development activities on PTR systems indicate that a pulse tube cooler design incorporating a low-temperature switch (LTS) valve is most ideal for industrial applications that require high refrigeration powers. A PTR design configuration uses a recuperative heat exchanger instead of a regenerative heat exchanger, and a low-temperature switch valve is installed at the cold end of the orifice pulse tube. Regenerative heat exchangers are best suited for small cryocoolers, where low to moderate cooling powers are required.

The adiabatic expansion efficiency of the orifice-based pulse tube incorporating an LTS valve is better than 40%. In other words, deployment of an LTS valve leads to an improvement in the adiabatic expansion efficiency. Higher adiabatic expansion efficiency is critical in the refrigera-

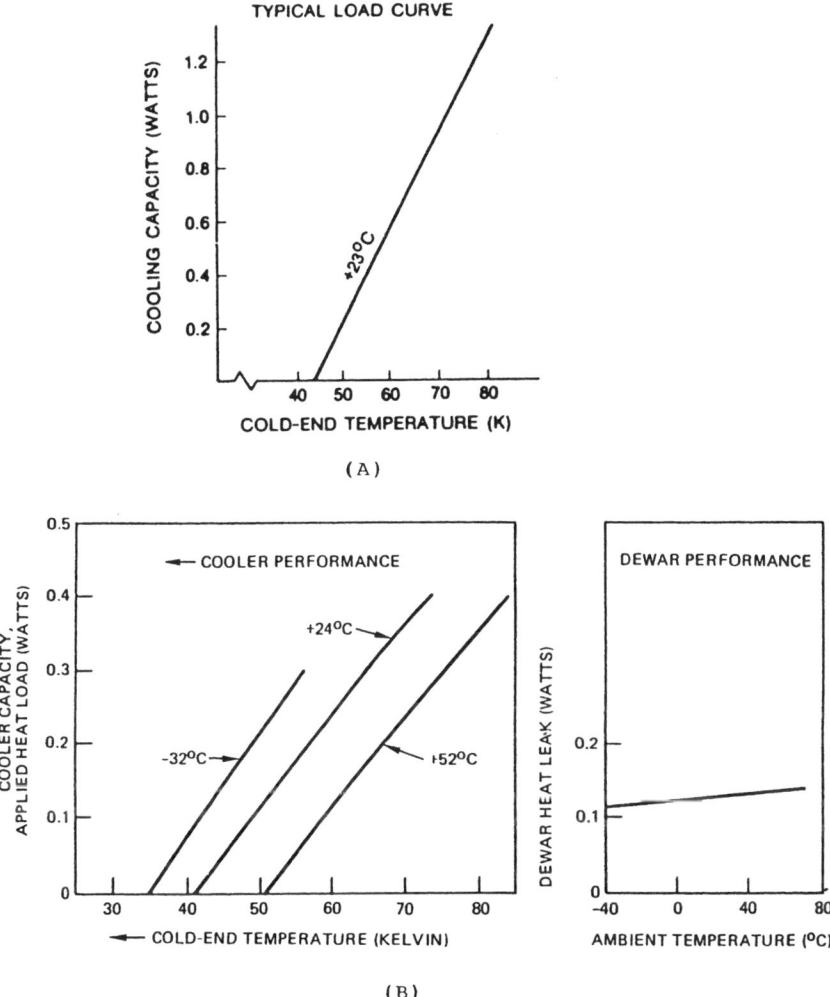

Fig. 7.8 (A) Cooling as a function of cold-end temperature and (B) cooler capacity and Dewar heat leakage as a function of ambient temperature.

tor systems that produce cooling powers on the order of kilowatts using mechanical expanders such as turbine expanders. A PTR system with an LTS valve is composed of a recuperative heat exchanger, gas-liquid separator, switching valve, gas flow straightener device, pulse tube, hot and cold heat exchangers, orifice, and gas reservoir [3].

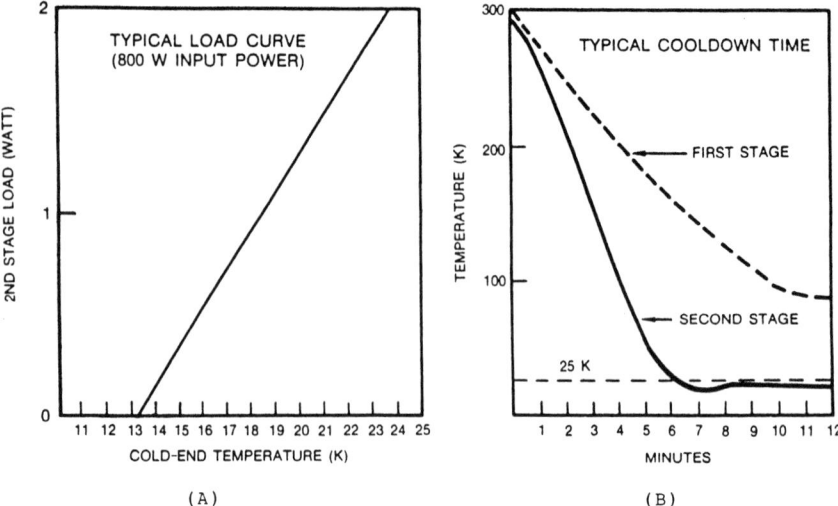

Fig. 7.9 Second-stage performance of a 0.25-watt miniature cryocooler (A) and cooldown time (B) for the first and second stage in the miniature 0.25-watt split Stirling cooler.

7.4.1 COOLING POWER AND ADIABATIC EFFICIENCY COMPUTATIONS

The cooling power (Pc) and adiabatic expansion efficiency (η_a) of industrial PTR systems are of paramount importance. The amount of heat (Q) at the hot end of the pulse tube is equal to the enthalpy drop at the cold end. Hence, the cooling power or cooling capacity can be written as

$$Pc = [dQ/dt] = f[H_1 - H_2] = [(dm/dt)C_p(T_1 - T_2)] \qquad (7.9)$$

where f is the operating frequency with a typical value between 1 and 3, H is the enthalpy of the gas during the expansion process, T is the temperature, C_p is the specific heat of the gas at constant pressure, dm/dt is the mass flow rate for gas, and subscripts 1 and 2 indicate relevant quantities before and after the adiabatic expansion process in the pulse tube. Assuming appropriate values for the parameters involved, computed values of cooling power as a function of temperature difference and pressure are summarized in Table 7.1. The computed values indicate that higher cooling power for commercial or industrial applications is only possible with helium gas at pressure ratios (pressure before expansion to pressure after expansion) exceeding 10.

Table 7.1 Cooling Power as a Function of Gas Temperatures and Pressure Ratio during the Expansion Process (Watts).

T_1-T_2 (K)	Pressure ratio = 1	Pressure ratio = 10
40	119	470
50	147	585
60	178	708
70	207	824
80	237	943
90	267	1063
100	296	1171

Table 7.2 Adiabatic Efficiency as a Function of Pressure Ratio and Outlet Temperature.

Inlet pressure (psi)	outlet pressure (psi)	pressure ratio	T_2 (K)	η_a (%)
72.5	14.5	5	260	41
101.5	14.5	7	257	34
130.5	14.5	9	253	31
159.5	14.5	11	245	38

The adiabatic expansion efficiency can be expressed as

$$\eta_a = [T_1 - T_2]/[T_1 - T_{2a}] \tag{7.10}$$

where T_{2a} is the temperature of the gas after adiabatic expansion, T_1 is the inlet gas temperature (which is close to ambient or room temperature), and T_2 is the outlet gas temperature in Kelvin. The temperatures T_2 and T_{2a} are a function of inlet and outlet gas pressures. Note that pressure P_1 (inlet pressure) is greater than pressure P_2 (outlet pressure). The computed values shown in Table 7.2 indicate that the adiabatic expansion efficiency is strictly a function of pressure ratio and the outlet gas temperature T_2 for a given orifice opening.

It is important to mention that the adiabatic efficiency attains the maximum value when the orifice opening is about 50% and the operating frequency is 2.6 Hz. In brief, adiabatic efficiency is dependent on pressure and temperature before and after expansion, actual temperature drop, adiabatic temperature drop, pulse tube diameter, orifice opening, operating frequency, type and size of regenerative materials, and pulse tube dimensions. The adiabatic efficiency of an industrial pulse tube cooling system with an LTS valve as a function of operating frequency, pressure ratio, and orifice opening is illustrated in Figure 7.10.

The temperature distribution along the longitudinal axis for a 40-cm long pulse tube with its orifice completely closed is shown in Figure 7.11(A), while the outlet gas temperature with and without a straightener device as a function of the pressure ratio is shown in Figure 7.11(B). Note that lower T_2 temperatures are possible with gas flow at low-pressure ratios when straighteners are used. The cooler efficiency is also dependent on the heat transfer capacity of the hot-end heat exchanger. Tradeoff studies performed by the author on PTR technology indicate that a conventional PTR system is inadequate for applications such as gas separation and liquefaction requiring large refrigeration powers. However, a PTR system when integrated with a low-temperature switch valve can meet the refrigeration power level requirements for industrial applications.

7.5 Multibypass and Active Buffer-Stage Techniques to Improve PTR Cooling Efficiency and Capacity

The cooling efficiency of PTR systems with low cooling power levels can be significantly improved by incorporating multibypass technique and two or more buffers connected at the hot end of the pulse tube through ON/OFF valves as illustrated in Figure 7.12.

7.5.1 IMPLEMENTATION OF ACTIVE-BUFFER STAGES FOR EFFICIENCY ENHANCEMENT

Implementation of active buffer stages in a PTR system decreases the loss through the high-pressure valve and low-pressure valve. These active buffers are integrated at the hot end of the pulse tube through ON/OFF valves [4] as shown in Figure 7.12. The active buffer offers a function similar to the tube expander. This type of PTR configuration has demonstrated a cooling capacity of 166 W at a refrigeration temperature of 80 K, an operating frequency of 2 Hz, high pressure of 2.2 Mpa (319 psi), and low pressure of 1.2 Mpa (174 psi). During the expansion cycle, work done (W) is dependent on enthalpy flow through the pulse tube at pressure P,

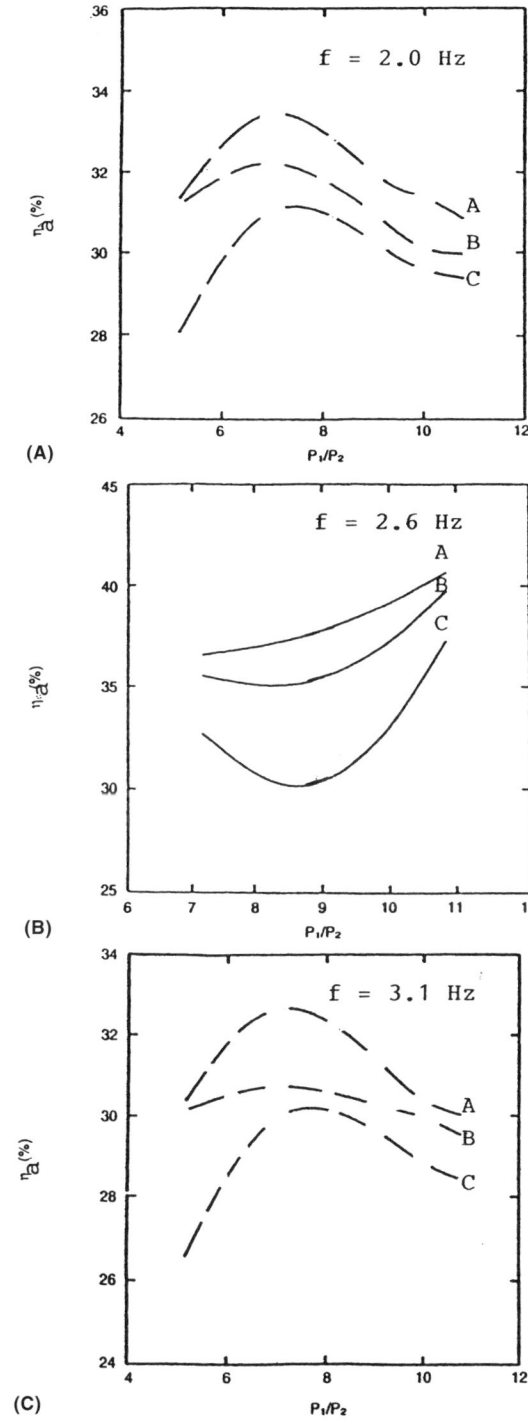

Fig. 7.10 Adiabatic efficiency (η_a) of a PT cooler with low-temperature switch valve as a function of pressure ratio and valve opening. Symbols: A = 50% valve opening; B = 100% valve opening; C = valve closed; and f = operating frequency.

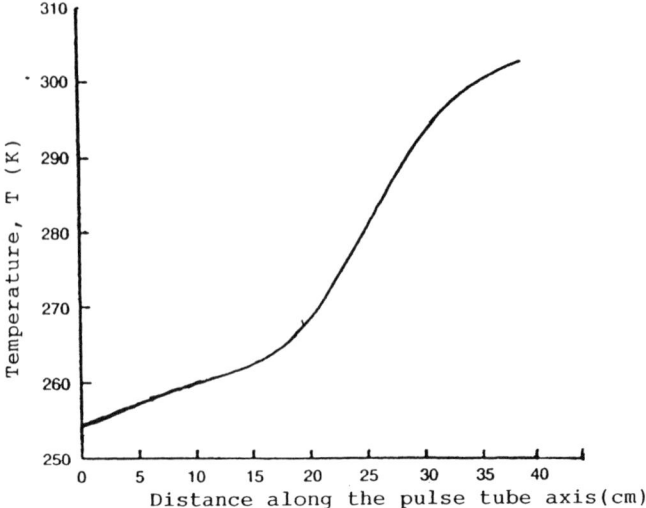

(A) Temperature distribution along the longitudinal axis with orifice completely closed [3]

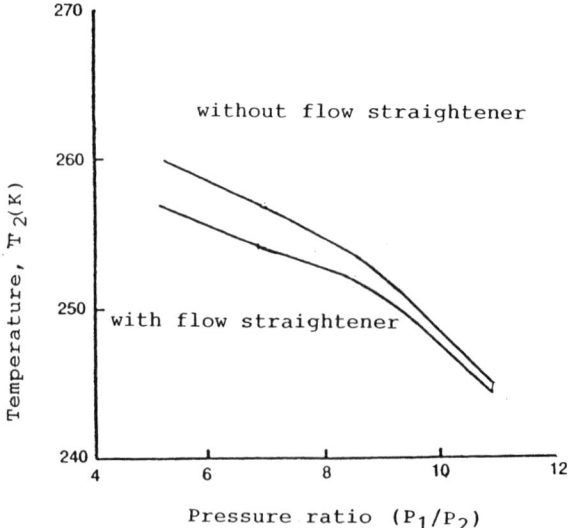

(B) Impact of gas flow straightener and pressure ratio on outlet temperature T_2

Fig. 7.11 Temperature distribution (A) and outlet temperature under various flow conditions and operating pressures (B).

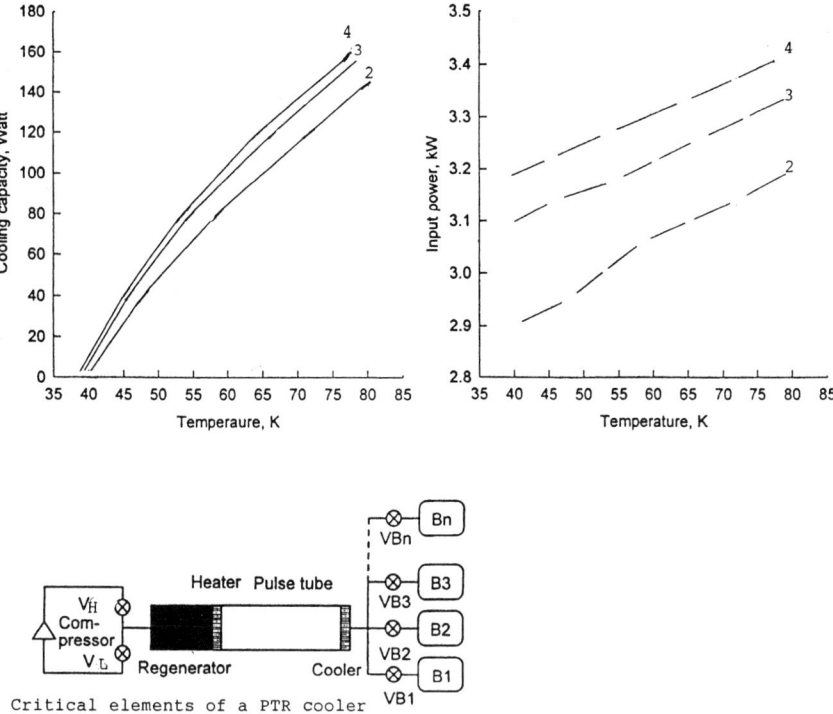

Fig. 7.12 Critical elements of a PTR cooler and cooling capacity for stage number 4, number 3, and number 2. Symbols: B_1, B_2, B_3 = buffer stage number 1, number 2, and number 3, respectively; V_{B1}, V_{B2}, V_{B3} = valve for buffer stage number 1, number 2, and number 3, respectively; and V_H and V_L = high and low pressure valves in the compressor unit.

enthalpy flow through the regenerator (H_R) as a function of pressure ratio (P_1/P_2) in the pulse tube, and cooling capacity (Q_c). P_1 and P_2 represent the pressure before and after the expansion, respectively. The preceding thermodynamic quantities increase almost linearly with the increase of the pressure ratio. The cooling capacity of a low power PTR system varies from 50 W at a pressure ratio of 1.3, to 75 W at a pressure ratio of 1.4, to 125 W at a pressure ratio of 1.5, to 155 W at a pressure ratio of 1.6 — all at an operating frequency of 2 Hz. Under these operating conditions, the mass flow rate from the active buffers is about 40% of that from the compressor, thereby realizing significant energy savings leading to higher cooling efficiency.

7.5.2 INTEGRATION OF MULTIBYPASS TECHNIQUE TO IMPROVE COOLING EFFICIENCY

Integration of multibypass technology in a PTR system (Fig. 7.13) offers remarkable improvement in reliability, longevity, efficiency, and vibration-free operation. When a bypass tube and a valve are connected to the middle of the pulse tube and the regenerator section, a part of the gas flows in or out of the pulse tube. The middle bypass technique has a positive impact on the pressure and mass flow rate at the cold head of the pulse tube. In a middle-bypass PTR system, all quantities including the enthalpy flow in the regenerator region through the middle-bypass valve, the enthalpy flow in the pulse tube, and the heat flow in the cold head undergo changes. Note that the "double-inlet" (DI) tube shown in Figure 7.13 acts

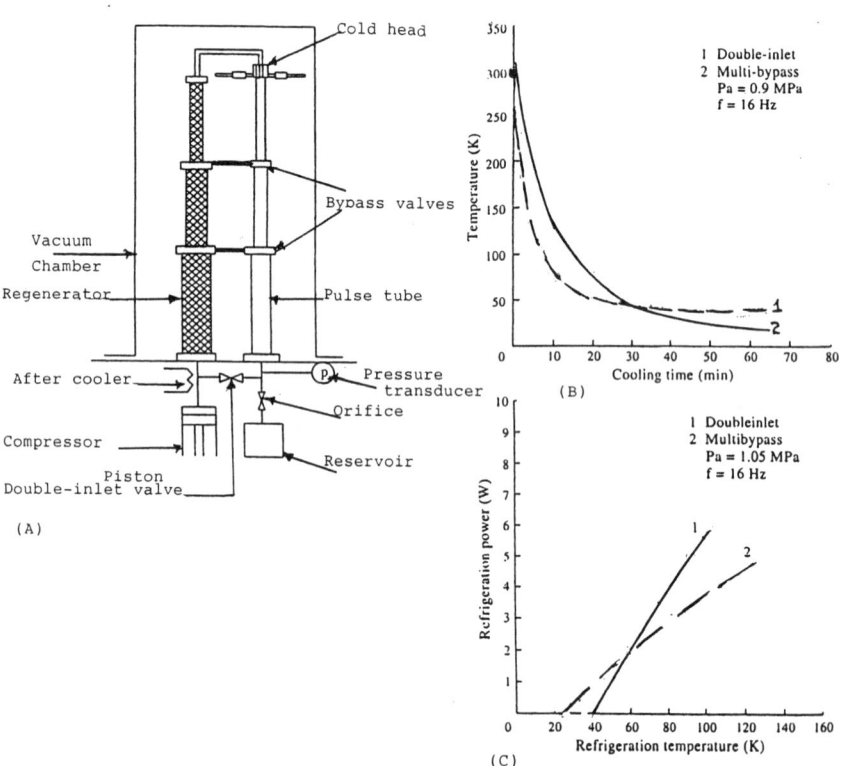

Fig. 7.13 Performance capabilities of a multibypass pulse tube: (A) critical elements of the MPPT, (B) cooling time comparison, and (C) refrigeration power as a function of cryogenic temperature.

as an expander in the refrigeration process. The DI tube increases the pressure ratio in the PTR, while the gas quantity decreases. The principal function of the DI tube is to adjust the phase shift between the pressure wave and mass flow at the hot head of the PTR and to increase their magnitudes.

Introduction of a multibypass tube between the middle of the pulse tube and the regenerator allows a part of the gas stream to flow into the PTR directly from the regenerator. This technique increases the pressure ratio and gas piston stroke in the cold section of the PTR, thereby achieving a low temperature in the shortest time. Integration of both the multibypass valve and DI techniques into a PTR system offers significant improvement in terms of lower cryogenic temperatures and higher cooling capacity. This kind of PTR system is best suited for space and airborne applications because of the benefits stated earlier. Note that refrigeration temperature increases with the increasing flow area of the multibypass valve. The multibypass-based PTR is essentially a kind of multistage PTR system, which produces multistage expansion processes: one at the cold head and the other at the multibypass port. The upper-stage cooling power is used to compensate for the losses occurring in the lower stage of the PTR system, resulting in a lower refrigeration temperature and higher cooling efficiency.

7.6 Cryocooler Requirements for Microwave, MM-Wave, and High-Power Laser Systems

This section defines cryocooler requirements for microwave systems, MM-wave sensors, infrared devices, and high-power laser systems with a major emphasis on reliability, cooling capacity, and refrigeration efficiency.

7.6.1 COOLER REQUIREMENTS FOR MICROWAVE AND MM-WAVE SYSTEMS

Significant performance improvement in microwave and MM-wave systems is most desirable in terms of noise figure, resolution, and operating range. Cryogenic cooling requirements for these sensors must take into account various environmental effects and mechanical tolerances. The physical dimensions of the cryogenically cooled low-noise microwave and MM-wave imaging systems must be such that their tolerances can be reasonably maintained under cryogenic operations over the 4–77 K operating range. Logistic and design problems may arise in cooling a critical RF component such as a parametric amplifier in a high-performance radar or high-resolution MM-wave imaging sensor. The most common problems that need to be overcome are moisture condensation, freezing of waveguide transmission lines, dimensional changes with temperatures, and variation in thermal conductivity as a function of temperature for various materials used in the fabrication. The effects of moisture condensation can be eliminated in two ways: (1) by employing the "sealed" system shown in Figure 7.14, where a moisture-free heat exchanger gas or partial vacuum is present, or (2) by using a moisture-free coolant gas to flow through the waveguide.

Note that a "sealed" system requires high hermetic sealing integrity of the microwave circuitry throughout the thermal cycling. Furthermore, the required performance levels of all the critical discrete components or devices must be maintained constant, regardless of the operating temperature. The critical components of the sealed microwave system [5] include the Dewar flask made from INVAR material, the waveguide window made from high-performance fused quartz, the positive pressure safety valve, a low thermal conductivity convection region, and an enclosure made from materials with a low coefficient of thermal expansion and high mechanical integrity. Cryogenically cooled microwave or MM-wave systems designed to operate at liquid nitrogen temperature (77 K) or liquid helium temperature (4.2 K) must address all radio frequency, mechanical, and thermal issues that can have an adverse impact on system performance under cryogenic operations.

7.6.2 *CRYOGENIC COOLER REQUIREMENTS FOR INFRARED DEVICES AND SENSORS*

Low to moderate cooling powers are required to cool IR detectors, IR imaging sensors, and IR missile receivers. Cryocoolers for these devices and sensors must have high reliability, improved refrigeration efficiency, and compact size. The Stirling cycle-based cryocooler shown in Figure 7.14(B) is best suited for spaced-based sensors, airborne imaging systems, IR detectors, and focal planar array (FPA) applications. A typical cryogenically cooled FPA includes the monolithically integrated 128X128 InSi-Schottky-Barrier detectors and a CCD readout circuitry on the same

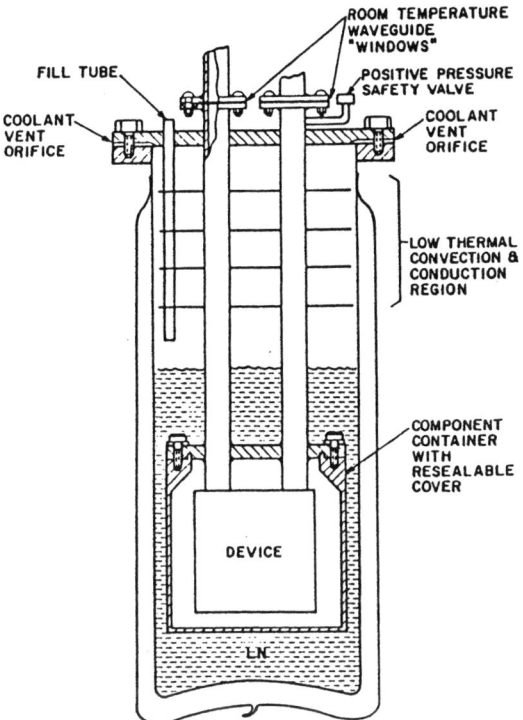

(A) A "sealed" liquid-nitrogen cooler for cooling the front end of radars

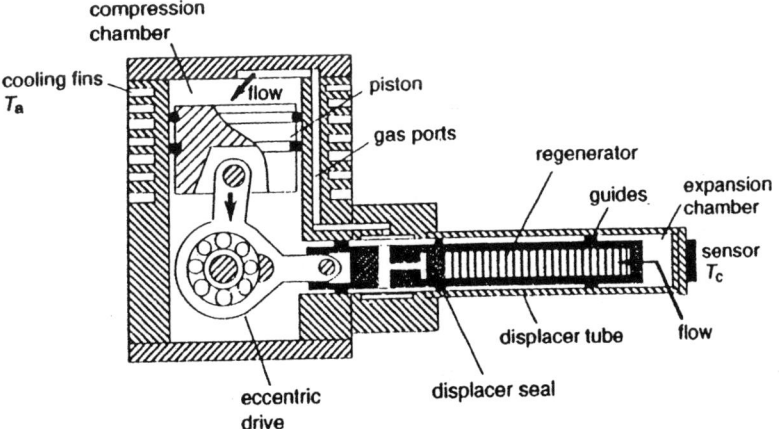

(B) Stirling-cycle microcooler for IR sensors and imaging systems

Fig. 7.14 Refrigeration systems for microwave radars and IR sensors.

chip. This kind of FPA when cooled to a temperature of 77 K has demonstrated high-quality imagery with a minimum resolvable temperature (MRT) better than 0.3 K over the long-wavelength infrared (LWIR) spectral region (8 to 14 microns). The MRT can be further improved to 0.12 K at 20 K and 0.05 K at 4.2 K cryogenic operations. The IR sensor, which is attached to the displacer tube, can be cooled down to 75 K temperature in a very short time with a refrigeration power of less than 1 watt. A sealed displacer package prevents the gas leakage. The displacer and the piston are driven sinusoidally through the eccentric drive mechanism. The 90-degree phase angle between the displacer and the piston produces the gas flow through the regenerator and the production of refrigeration in the expansion chamber. The refrigerator generally deploys small lead spheres or low-cost rare-earth materials that are tightly packed together and held on either end of the regenerator. The displacer for these cryocoolers is made from low-thermal-conductivity materials such as glass epoxy composite. The regenerator is designed into the piston to eliminate excessive heat losses. Minimizing thermal losses and maximizing the thermal performance of the regenerator in a cryocooler are critical. It is important to point out that the regenerator's performance depends on the mass of the working fluid passing through the regenerator during the cooling period, the fluid mass cooling of the thermal load in the heat exchanger, and the average specific heat of the working fluid. The average specific heat (Cp) of the helium varies over the temperature range from 300 to 75 K. However, the average specific heat of the regenerator section is equivalent to the specific heat of the gas in the heat exchanger and the ratio of the available refrigeration to the heat transferred between the streams in the regenerator. This particular ratio is close to 100:1 for a cryogenic cooler producing 1 W of refrigeration power with an expansion volume of 1/3 cm^3. This means that the transferred heat in the regenerator is two orders of magnitude greater than the available refrigeration produced by the cryocooler.

To limit the refrigeration loss to a minimum, the heat exchange efficiency must exceed 97% to achieve a usable refrigeration capacity in a small cryocooler, and the efficiency of the regenerator must be well above 99%. This kind of microcooler design uses a heat exchanger with a size equivalent to a cigarette package that is capable of exchanging heat of 100 watts between the streams over a temperature difference of 225 K with less than 1 watt of heat leaving the regenerator and leaving the refrigeration space. This microcooler has potential applications in airborne missiles and space sensors, where weight, size, and power consumption are the critical requirements. A microcooler designer must have a thorough understanding of the thermodynamic properties of working fluids and materials used, fluid dynamic principle, and the latest material technology.

A Stirling-closed cycle microcooler designed for a NASA thermal imaging radiometer when operated at optimum cryogenic temperature demonstrated thermal images with high quality and resolution. This Stirling-closed cycle microcooler, developed by INFRAMETRICS Inc. of Billerica, Massachusetts, weighs less than 11 oz, consumes electrical power of less than 3.5 W with 150 mw heat load, has an operating life that exceeds 2000 hours of continuous operation, and has a cooldown time for 100 Joule mass of 1.5 minute to 120 K and 4 minutes to 77 K from room temperature of 300 K. Since the microcooler eliminates the need for liquid nitrogen, battery-powered portable cooling systems can be developed and marketed for commercial applications. In summary, the closed-cycle Stirling coolers are compact, lightweight, require minimal electrical power, and, therefore, should find many applications in commercial, industrial, space, and military systems such as IR search and tracking (IRST), IR line scanner (IRLS), IR countermeasures (IRCM), and IR missile warning (IRMW) receiver.

As stated earlier, leading edge detector technology such as FPA technology is best suited for important military IR sensors and devices. For example, an IRST, IRLS, or IRMW receiver would prefer the integration of cryogenically cooled focal planar arrays (FPAs) for optimum sensitivity and reliable sensor performance. Cryogenic cooling of the FPAs used in IR receivers and analog-to-digital converters (ADCs) employed in a digital signal processing (DSP) unit is absolutely necessary if high system sensitivity, improved resolution, and high computational capability are the principal requirements. The detectivity of an FPA depends on cryogenic temperature and processing gain offered by the detector elements used by the FPA. It is important to mention that the detectivity of IR detectors and FPAs significantly improves at lower cryogenic operations (4.2 K). Hg:Cd:Te detectors, when cooled down to 77 K, provide large dynamic range, improved sensitivity, and optimum spectral response over the 2- to 12-micron range. Stirling coolers are widely used to cool IR detectors, to reduce their dark current levels, and to improve the detectivity. IR detectors made from the Hg:Cd:Te compound have demonstrated a detectivity greater than 10^{11} cm$\sqrt{\text{Hz}}$/W at cryogenic operating temperatures below

77 K, which is 10 times better than that at room temperature (300 K). The detectivity could be further improved close to 10^{12} cm√Hz/W if the operating temperature is reduced to 4.2 K. Cryogenic cooling of both the FPA and time delay integrating (TDI) circuits will significantly improve the detection and tracking capabilities of IR sensors.

7.6.3 REFRIGERATOR REQUIREMENTS FOR HIGH-POWER LASER SYSTEMS

Cryogenic cooling is absolutely necessary to maintain the efficient and safe operation of high power laser systems used in industrial, medical, space, and military operations. An optical window located at the output of the laser transmitter is the most sensitive and critical component of a high-power laser system, regardless of applications. Cryogenic cooling at the right temperature is necessary to avoid the catastrophic failure of a high-power laser system. High-power laser systems include gas lasers such as CO_2–TEA operating at 10.6 microns and chemical lasers such as hydrogen fluoride (HF) lasers operating at 2.744 microns and deuterium fluoride (DF) lasers operating at 3.875 microns. The high-power lasers have potential applications in industrial, medical, and military fields. These lasers offer maximum power levels at operating wavelengths as specified previously. High-power TEA lasers have demonstrated peak power levels exceeding 1 gigawatt (1000 MW) at 10.6 microns with conversion efficiencies better than 10%, while the chemical lasers (DF and HF) have delivered the highest continuous wave (CW) power levels, close to 100 kW. Low-power CO_2 lasers with a CW power level of around 10 watts and chemical lasers DF and HF lasers with 10 CW power of 10 watts or so are best suited for scientific research and laser range finding applications.

As stated earlier, optical windows are the most critical elements of high-power laser systems. Cryogenic cooling of the optical windows is absolutely essential to maintain desired beam quality, laser output power level, and stated reliability over extended durations. The selection of window material and cryogenic operating temperature depends on the laser CW output power and period of continuous operation. High-power laser systems operating at 10.6 microns for long-range missile tracking and detection are capable of providing output peak energy in excess of 300 joules, maximum power greater than 3 GW, conversion efficiency better than 10% depending on the CW power output, and brightness exceeding 4×10^{15} W/cm^2/Sr. These high-power lasers require cryogenic coolers with ultrahigh cooling capacities to maintain the output window temperature at a specified operating temperature. A high power laser has potential application in an airborne system capable of detecting, tracking, and destroying long-range hostile missiles during its launch and boost phases. Keeping the optical window of this laser at the right cryogenic temperature is very important in such critical military operations.

The latest laser technology reveals that a COIL operating in the lower IR spectral region around 1.315 microns is best suited for specific commercial and military applications. A COIL laser radiates power at a single laser line at 1.315 microns, generates a high-quality beam, reduces turbulent effects in laser cavity, lowers differential effects in the atmosphere, possesses good thermal properties, and offers conversion efficiency as high as 25% depending on the power output level.

A COIL laser has demonstrated the ability to cut steel plates and concrete structures with thicknesses up to 12 inches into small pieces, even with a CW power less than 10 kW. The cryogenic coolers operating at 77 K with moderate cooling capacities are adequate for cutting applications. This type of laser is best suited for commercial and industrial applications involving dismantling decommissioned nuclear power plants. Because of its unique performance capabilities, the COIL laser can play an important role in the National Missile Defense (NMD) system for protection from long-range missile attacks.

7.7 Cryogenic Coolers for Sonar Applications

An acoustic sonar transmitter using high-temperature superconductor (HTSC) technology was developed under Small Business Innovative Research (SBIR) contract [6]. This particular transmitter system employs three distinct technologies: HTSC technology, magnetostrictive technology, and cryocooler technology. The system is composed of magnetostrictive elements, a pair of magnetic coils wound using HTSC wire, and a Stirling-cycle refrigerator to maintain a cryogenic temperature of 50 K. The cooling provided by the Stirling cooler is reliable and effective. This type of cooler can provide cooling power in excess of 250 watts at 77 K. The same cooler can be operated over a 50 to 55 K temperature range, if required, with cooling time of less than 4 hours. This high-power

Stirling-type cooler is acoustically quiet and has an impressive reliability record, better than 50,000 hours, and, therefore, is best suited for submarine sonar applications. High-power cooling at 25 K and at low frequencies are needed for accurate mapping and detection of underwater targets at long ranges, but at the expense of high cost and complexity. Furthermore, this particular cooler can operate successfully in all directions and does not require a periodic replenishment of liquid cryogen. The cooler is best suited for ship propulsion, levitated trains, and superconducting generators and motors. This cryogenically cooled sonar transmitter provides seismic topography, global warming measurements, ocean floor mapping, ocean current monitoring, and underwater target detection with high precision, minimum acoustic signatures, and improved reliability.

7.8 Cryogenic Coolers for Medical Applications

Cryogenic coolers are widely used for medical diagnostic equipment including magnetic resonance imaging (MRI), computer tomography (CT), CT scanners, magnetic cardiogram mapping sensors, superconducting quantum interference device (SQUID)-based gradiometers for measurements of neurological signals, second-order gradiometers for neuromagnetic research, and multichannel SQUID-based magnetometers for accurate measurement of three components of heart magnetic vectors to detect heart diseases. Note that DC-SQUID-based magnetometers operating at 4.2 K provide the optimum sensitivity and resolution needed for reliable medical diagnosis. The multichannel neuromagnetic gradiometers when cooled down below 4.2 K temperature offer an intrinsic noise level of less than $20 fT/\sqrt{Hz}$ down to a frequency of 0.5 Hz without the use of electronic noise cancellation technique. When operated below a temperature of 4.2 K, this sensor offers vital information to brain surgeons prior to undertaking a complex and risky surgery because it simultaneously measures five critical magnetic field components surrounding the tumor in the brain. A compact DC-SQUID magnetometer when integrated with a 4.2 K cryocooler acts like portable diagnostic equipment capable of recording the tangential components of the magnetic fields associated with the rhythmic alpha activity in the brain. Accurate measurement of several components of magnetic fields in the brain and heart is necessary for their reliable clinical assessment and functional evaluation. Vibration-free cryogenic coolers operating at 4.2 K or lower temperatures are best suited for medical diagnostic equipment to provide improved sensitivity, high accuracy, enhanced reliability, and optimum patient comfort.

7.9 Summary

Implementation of the latest technologies and materials in the cryocooler design is necessary for enhanced reliability, higher adiabatic efficiency, and improved cooling capacity. This chapter briefly discussed the performance requirements of cryocoolers best suited for industrial, scientific research, military, space, and medical applications, with a major emphasis on cooling power level and refrigeration efficiency. Critical performance parameters of G-M-cycle and Stirling-cycle coolers for commercial and military applications were summarized with emphasis on cooling efficiency and cooldown time. Mathematical expressions for cooling power and adiabatic expansion efficiency for PTR systems were provided as a function of refrigeration temperature, pressure ratio, and operating frequency. Computed values of cooling power and adiabatic efficiency as a function of cryogenic temperature, orifice opening, operating frequency, outlet gas temperature, and gas pressure ratio were summarized for various coolers. Techniques such as the active-buffer concept, the multibypass method, and implementation of high-performance regenerative rare-earth materials to improve cooling capacity and refrigeration efficiency were described. Refrigerator system requirements for microwave and MM-wave systems, infrared sensors and devices, high-power laser systems for commercial and military applications, and sonar transmitters for underwater target detection and surveillance were identified. Cooling system requirements for high-power airborne laser systems to detect and track long-range missiles were summarized. Cryocooler types and their performance requirements for medical diagnosis, clinical assessment, and functional evaluation were identified, with a major emphasis on sensitivity, reliability, resolution, and patient comfort.

References

1. A. Bittermann. "Foothold in cryogenic materials." *Superconductor Cryogenics*, Spring 1998, pp. 12–17.
2. G. Chen et al. "Studies of the minimum refrigeration temperature of regenerative cryocoolers." *Cryogenics*, 1977, vol. 37, no. 7, p. 399.
3. J. Liang et al. "Pulse tube refrigerator with low temperature switch valve." *Cryogenics*, 1997, vol. 37, no. 8, pp. 397–398.
4. S. Zhu et al. "Investigation of active-buffer pulse tube refrigerator." *Cryogenics* 1997, vol. 37, no. 8, p. 462.
5. T. P. Duprex. "Designing cryogenic microwave systems." *Microwaves*, October 1967, pp. 39–41.
6. C. H. Joshi et al. "Putting chill into HTSC applications." *Superconductor Industry*, Fall 1993, pp. 30–35.

Chapter 8 | Requirements for Cryogenic Materials and Associated Accessories Needed for Various Cryogenic Coolers

8.0 Introduction

This chapter defines the requirements for the cryogenic materials, cryo-electronics, and associated accessories needed to maintain the required performance of a cryogenic refrigeration system without compromising cooler reliability and cooling efficiency. The integration of the next generation of cooler technology, which could lead to higher heat load capability and improved reliability is discussed. Dilution refrigerator systems are best suited to provide cooling temperatures as low as 2.9 mK. In applications where quasi-cryogenic or higher cryogenic temperatures are required, heat pipes, mechanical refrigerators, magnetic refrigerators, and thermoelectric (TE) coolers can provide cooling beyond the 150 to 270 K range with a minimum of cost and complexity. Integration of the heat pipe technology along with surface mount technique in solid-state devices offers the most cost-effective approach for removing the heat dissipated at higher temperatures. Performance requirements for the associated accessories, auxiliary equipment, and refrigerants are summarized. Important characteristics of potential coolants or cryogens are briefly summarized, with an emphasis on cooling capacity and operating temperatures. Maintenance requirements, frequency of service, and field maintenance procedures are outlined wherever possible. Operational problems frequently encountered in cryogenic refrigeration systems such as moisture condensation, heat loss, freezing of lines carrying the coolants, radical changes in the component dimensions at cryogenic temperatures, variation in thermal conductivity with temperature, and leakage of the refrigerant are identified. Techniques to overcome the operational problems and to minimize their impact on the cryocooler performance and reliability are summarized, with an emphasis on maintenance procedures. Cryogenic leak in tube fittings and cryogenic insulation requirements are also briefly discussed.

8.1 Cryocooler Requirements for Space-Based Communications and Surveillance Imaging

Space-based communications systems and surveillance imaging sensors involve various devices such as radio frequency/microwave, MM-wave, digital, optoelectronic, Josephson Junction (JJ), and the superconducting quantum interference device (SQUID). It is important to mention that these systems must deploy space-qualified microcooler packaging and interconnects, closed-cycle cryocoolers, and cryoelectronic interface devices to meet functionality, reliability, and survivability requirements. Studies performed by the author indicate that compact, efficient Stirling and state-of-the-pulse tube cryocoolers are best suited for space-based or satellite-based sensors. These cryocoolers are equipped with in-flight control electronics, which provide a thermal and mechanical interface between the system and the cryocooler and magnetic shielding to reduce the magnetic noise produced by the cryocooler motor well below JJ and SQUID devices.

8.1.1 COOLER REQUIREMENTS FOR FRONT-END COMPONENTS

The front-end components are housed in the cryocooler enclosure made from high permeability magnetic material. The cryogenic-cooler package must be tested under thermal vacuum and flight vibration conditions. Active electronic control is required to compensate for the expected temperature variations in the heat rejection plate, which allows power dissipation in the payload. The cryocooler for this particular application must be subjected to full space qualification tests, including eletromagnetic interference (EMI), vibration, and thermal vacuum tests, prior to installation in the payload.

Cryogenic cooling of the front-end components is essential in satellite communications and terrestrial mobile communications systems to reduce the noise in the front end amplifier and insertion loss in band select filters, multiplexers, delay, lines, and analog-to-digital (A/D) devices. Note that a reduction in the noise figure offers increased channel capacity or more

8. Requirements for Cryogenic Materials and Associated Accessories

Table 8.1 Mass of Controller and State-of-the Art Space-Qualified Cryocooler.

Cooling power (W) (at 77 K)	Mass of the cryocooler (Kg)	Mass of electronic control (Kg)
4 W	1.25	2
10 W	5.00	4
30 W	12.00	5

simultaneous users per area. Cryocoolers for satellite and terrestrial communications must provide coefficient of performance (COP) better than 5% at 77 K, maintenance-free life exceeding 5 years, and operational capability under a wide range of thermal environments. State-of-the art space-qualified Stirling-type coolers using linear compressors and pulse-tube cryocoolers are best suited for space and tactical applications. Studies performed by the author reveal that a superconducting multiplexer when integrated with pulse-tube cryocoolers operating at 77 K represents the potential of performance enhancements of strategic value to future military sensors, satellite communications systems, and high-resolution surveillance and imaging equipment.

Direct current (DC) input power to the cooler is proportional to the cooling load. In both applications, limited DC power is available for cryocoolers. It is important to mention that a space-qualified cooler efficiency at 77 K and at a rejection temperature of 300 K is about 6.6%, which means that 15 W of DC power is required for 1 W of cooling load. Table 8.1 summarizes the mass of state-of-the-art space qualified cryocoolers and electronic controllers as a function of cooling capacity at 77 K temperature.

Integration of high-temperature superconducting technology in electronic and microwave devices offers the potential of a tenfold miniaturization of payload electronic equipment leading to significant improvement in the noise figure, gain/temperture (G/T) parameter, and effective instantaneous radiated power (EIRP).

8.2 Cryocooler Requirements for Military Applications

Cryocooler requirements for radar, missiles, electronic warfare, remote sensing systems, airborne surveillance, and high-power lasers for track-

Table 8.2 Performance of RF and Optoelectronic Devices AT 77 K.

Device type	Performance parameters	Cryogenic temperature (K)	
		300	77
Photodetector	Dark current (mA)	1	0.4
	Bandwidth (GHz)	12	14
Laser diode	Bandwidth (GHz)	13	16
GaAs amplifier	Noise figure (dB)	6.4	3.6
	Bandwidth (GHz)	12	15

ing hostile missiles are very stringent, particularly on reliability, weight, and size. Cooling capacity, efficiency, reliability, and maintenance-free operation are the principal requirements for airborne military systems. In case of air-to-air or air-to-surface missiles equipped with RF or MM-wave or optical seekers, cooler weight, size, and minimum time to achieve the desired cryogenic temperature are the critical performance requirements.

Compressive receivers, widely used in electronic warfare applications, yttrium barium copper oxide (YBCO)-chirp filters with Hamming weightings, A/D devices, rapid-single-flux-quantum (RSFQ) chips, amplifiers, and detectors using advanced high-speed semiconductor technology are widely used in the receiver architecture. This type of compressive receiver is superior in overall size, weight, and power consumption when cooled at 77 K. Pulse-tube cryocoolers with linear compressors will meet the cooling capacity and efficiency for this receiver. Significant performance improvements in microwave and optoelectronic devices as illustrated in Table 8.2 are possible at 77 K. These devices will require maintenance-free closed-cycle cryogenic coolers with moderate cooling power. Furthermore, the coolers for military applications must meet high efficiency and reliability standards in addition to vibration-free operation.

8.2.1 TACTICAL COOLERS FOR INFRARED MISSILES

In the case of airborne IR missiles, the cryocooler is required to cool down the detectors and other optoelectronic components in the shortest possible time needed for a successful mission. The cryocooler for a tactical mission is called a tactical cooler and must have sufficient cooling capacity to achieve the required cryogenic temperature in the shortest possible time frame. Cooler maintenance for this application is not critical. Studies performed by the author indicate that a linear-type Stirling cooler with a cooling power of 3.5 W at 77 K is best suited for IR missiles. The latest version of the cooler is capable of delivering a nominal cooling power exceeding 10 W at 77 K. A pulse-tube with a cold finger can be designed to achieve a maximum cooling power of 10 W at 77 K and an operating life exceeding 3 years. In very sensitive applications, the pulse-tube cooler using two cold fingers cancels vibrations. However, a cooler with a single finger offers a less expensive and more efficient cooler configuration.

8.3 Dilution Refrigeration Systems for Scientific Research

Dilution refrigeration systems are best suited for applications where extremely low cryogenic temperatures are required. These systems are widely used in scientific research and biomedical studies. The performance of a dilution refrigerator depends on enthalpy, circulation rate, and applied power. The minimum achievable temperature is strictly dependent on the viscous heating efficiency, which takes into account the enthalpy coefficient of the helium and the heat exchanger. A well-designed dilution refrigerator system can provide a cooling temperature as low as 2.9 mK. This low temperature cannot be achieved from any other refrigerator system.

8.4 Cryocoolers for Higher Cryogenic Temperatures

For certain applications, higher cryogenic temperatures or quasi-cryogenic temperatures from 180 to 270 K are adequate to meet the performance requirements. Mechanical refrigerators, magnetic refrigerators, and thermoelectric (TE) coolers are best suited for operations where higher cryogenic temperatures are required. These cooling systems are briefly described next, with an emphasis on cost and complexity.

8.4.1 MECHANICAL REFRIGERATOR (MR)

Mechanical refrigerators are best suited for applications where operating temperatures are in the range of 180 to 225 K. The heat removable capacity of this particular refrigeration system is determined by considering three factors: the total amount of heat to be dissipated, the lowest temperature to be achieved, and the rate of temperature change in Kelvin per minute. An MR system has several advantages over an expandable refrigerant as a means to cool electronic devices in thermal cycling. The greatest single advantage of the MR is that it relies on the fixed amount of refrigerant that is installed at the factory and never needs replenishment. The MR refrigerator eliminates the complete storage and distribution system of the expendable refrigerants. Furthermore, no more tanks, no more insulated distribution lines, and no more bulky storage cylinders are required for the MR. The MR system only requires electrical power to operate and can be rolled to anywhere in the facility to provide quasi-cryogenic cooling. The MR system must be properly sized to meet higher cooling temperatures, whereas with expendable refrigerants there is a virtually inexhaustible supply of available heat removable capacity. However, MR systems are not suitable for lower operating temperatures.

8.4.2 MAGNETIC REFRIGERATOR SYSTEMS (MRS)

A magnetic refrigerator system consists of a superconducting magnet with magnetic field intensity close to 5 Tesla and is directly connected to the cooling chamber. The MRS operation is based on magnetocaloric effect, which occurs in magnetic materials known as ferromagnetic. These materials become hot when magnetized and become cold when demagnetized. Helium gas is used as a heat transfer fluid between the natural gas input and the solid magnetic refrigerant, which happens to be a gadolinium, a rare-earth ferromagnetic material. The gadolinium plays the role of coolant or refrigerant, staying solid throughout the cycle, while the strong magnet replaces the compressor. The gadolinium concentrates heat when a strong magnetic field of 5 Tesla is present and then quickly cools down to 270 K. In a 5-Tesla field, a cooling load of 600 W is achieved at an efficiency of 50%. Note that a conventional home refrigerator provides about 200 W of cooling at 30% efficiency. Magnetic refrigerators are best suited for applications such as air conditioning systems needed for conference halls and large office buildings.

8.4.3 TE COOLERS

TE coolers are best suited for applications where an individual device in a sensor requires cooling at subzero temperature. TE coolers are generally used to cool low-noise amplifiers, detectors, and charged coupled devices (CCDs). Studies performed by the author indicate that even cooling these devices at 178 K significantly improves the noise figure, sensitivity, and dark current. TE coolers have been used to achieve performance of CCDs with minimal cost and complexity. A sensor integrated with a TE cooler can be designed for a high-end spectroscopy and a high-resolution imaging camera. All critical components are housed in a ceramic Dewar with a permanent seal. In other words, a permanent hermetic vacuum is considered to control the already low level of outgassing. The TE coolers are placed very close to the CCD to achieve the lowest possible temperature of 178 K. Note that better cooling offers lower dark current and and fewer blemishes. The permanent hermetic vacuum ensures that the peak quantum efficiency as well as the cooling efficiency will not degrade, even after years of operation. Furthermore, elimination of condensation means that the system can use a single window with antireflection coating, which further protects against any degradation in peak quantum efficiency.

The TE cooling scheme has been recognized as most cost-effective when the operating temperatures are in the range of 150 to 270 K. TE cooling requires no moving parts. Simplicity, reliability, and economy are the outstanding features of a TE cooler. It is the most inexpensive cooling scheme. A single-stage TE cooler consists of a p- and n-type semiconductor compound connected with a low-loss metallic piece of wire or conductor. The operating principle of a TE cooler is based on the Peltier phenomenon, which states that when two dissimilar metals are connected in series with a specified electromagnetic force (EMF), one junction will be cooled and the other will be heated. The rate of heat pumping or removal depends on the magnitude of the current flow and the properties of the materials involved. Major advantages of a TE cooler include noise-free operation, cooler operation independent of orientation, low cost, and minimum size and weight.

8.4.3.1 Performance Parameters of TE Coolers

Critical performance parameters of a TE cooler include the figure of merit (FOM), which is dependent on the conducting material properties, the temperature difference, and COP. Published literature indicates that the average value of FOM for a p- and n-type bismuth-telluride alloy varies between 0.005 and 0.0005 per Kelvin [1]. TE coolers are best suited for cooling infrared detectors, focal planar arrays (FPAs), and other electronic devices or circuits requiring cooling temperatures between 150 to 200 K. Low-power, low-temperature TE coolers are most attractive for commercial and military applications where cost is the most critical requirement. A four-stage TE cooler [2] with a cooling temperature of 190 K and a cooling load of 10 mw requires electrical input power of about 1.7 W with a COP of 0.0059, whereas a six-stage TE cooler with an operating temperature of 170 K and a cooling load of 10 mw requires an input electrical power of 40 W yielding a COP of 0.00025. The COP parameters for both TE coolers are extremely low. More efficient TE coolers would require high-quality, low-temperature n- and p-semiconductor materials. Typical performance parameters of low-power, low-temperature TE coolers are shown in Figure 8.1. In summary, TE coolers provide cost-effective ways and means to remove excessive heat from small devices or circuits using the vapor-phase refrigeration technique. This particular cooler design suffers from extremely poor efficiency and is only recommended for low-power applications not exceeding a few tens of milliwatt of heat load.

8.5 Heat Pipe Concept for Higher Cryogenic Temperatures

In heat pipe technology, when treated as an integral part of a solid-state device such as a high-power RF amplifier or oscillator, the heat pipe arrays provide the most effective method for heat transfer from cold-plate, dissipating localized heat fluxes, eliminating hot spots, and reducing the maximum wafer temperature essential for high device reliability [3]. The widespread use of microheat pipes needs to address several critical issues such as the effect of the precise shape and size of the heat pipe, the effect of variations in the heat pipe array density, and projected performance degradation with respect to operating time. The latest research studies indicate that incorporating relatively large microheat pipes with diameters exceeding I mm or arrays of microheat pipes as an integral part of a silicon chip offers the most cost-effective cooling scheme for numerous heat transfer applications involving complex missile systems and space reconnaissance sensors.

8. Requirements for Cryogenic Materials and Associated Accessories

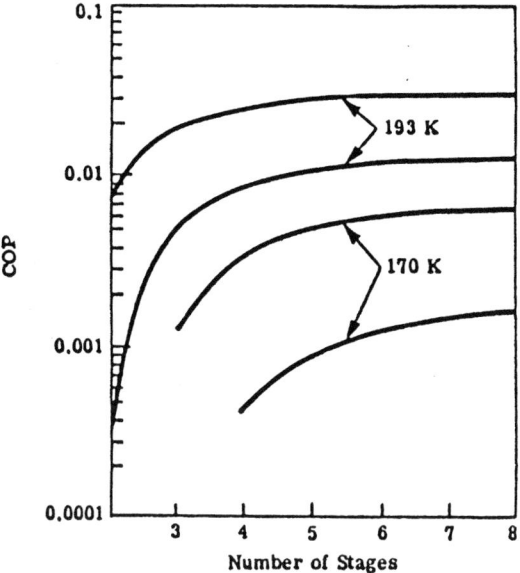

(A) Coefficient of performance (COP) for the bismuth-telluride p- and n-junction material

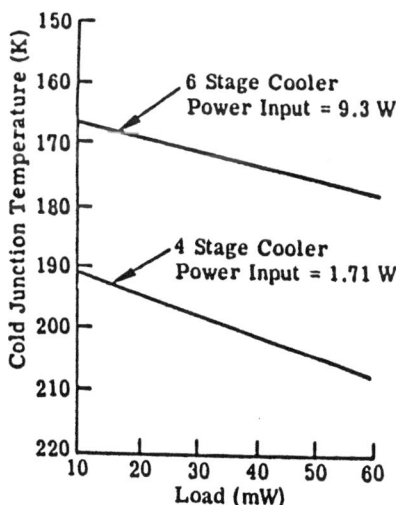

(B) Load capacity as a function of junction temperature

Fig. 8.1 Load capacity versus junction temperature (A) and heat load capacity for multistage TE coolers [1].

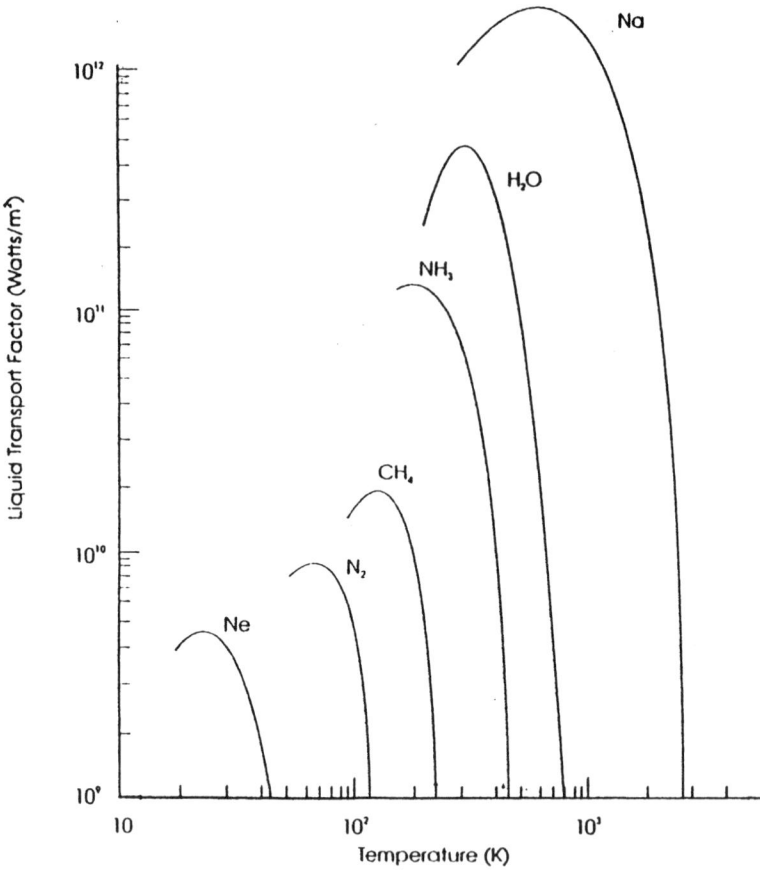

Fig. 8.2 Liquid transport factor or figure of merit (FOM) for various working fluids in heat pipes as a function of cryogenic temperature [3].

8.5.1 PERFORMANCE LIMITATIONS DUE TO WORKING FLUIDS IN HEAT PIPES

Cryogenic heat pipes employ either a chemically pure element such as helium, argon, krypton, or oxygen or a chemical compound such as methane, ethane, or freon as the working fluid. The operating temperature range and figure of merit (FOM) of these fluids vary from one working fluid to another. FOM is the leading performance measuring parameter, and its typical values for various working fluids are shown in Figure 8.2. New fluids are being developed to fill the gaps in the operating temperature ranges that exist between the critical point of helium (5.4 K) and the freezing point of hydrogen (13.8 K), and the critical point of neon (44.4 K) and the freezing point of fluorine (53.5 K).

It is important to mention that the use of cryogenic fluids in heat pipes presents some problems that are attributed to low surface tension, thermal conductivity, latent heat of vaporization, and high liquid viscosity present in the heat pipes. The most significant problem is due to surface tension under higher gravity (g)-environments.

Performance of heat pipes can be limited by the capillary limitation in which there is insufficient capillary pressure to pump the fluid back to the condenser, the sonic limitation in which the vapor velocity inside the heat pipe reaches the sonic condition [3] restricting the flow, the entrainment limitation in which the liquid drops are entrained in the vapor flow, and the boiling limitation in which high-vapor radial heat flux causes boiling in the evaporating wick. These limitations can be overcome using specific techniques. For example, capillary limitation can be overcome using improved wicking structures that increase the capillary pressure with a minimum of frictional pressure drop in the pipe. The boiling limitation is strictly dependent on the type of wicking structure, and the structure must be optimized for both normal and low-gravity environments. Homogenous wick structures such as screens, fibers, arterials, and axially grooved wicks are widely used [3] in cooling schemes using heat pipes.

The outstanding heat transfer capability of heat pipes make them most attractive for cooling embedded electronics in applications where the electronic circuitry must be enclosed and isolated from dust, moisture, oil mist, or harsh environmental conditions. An electronic system configuration composed of a heat exchanger with eight rows of 17 finned copper heat pipes is shown in Figure 8.3(A). The power level as a function of temperature drop is depicted in Figure 8.3(B), thereby demonstrating the dependency of the heat transfer capability of the heat exchanger. Implementing heat pipe technology in a system reduces the heat exchanger volume by 50% to 70% depending on the number of fins used.

A heat exchanger composed of two 25 mm-diameter copper heat pipes has demonstrated the thermal control of a high-power RF transistor pair (with an overall rating of 50 W) mounted on a cold plate with several fins. The RF power output level of 30 W was maintained with 500 grams of refrigerant or cooling agent in the heat pipe. This indicates that the heat pipe technology offers the most effective means for heat transfer from solid-state devices with minimum cost and complexity. Heat pipes must

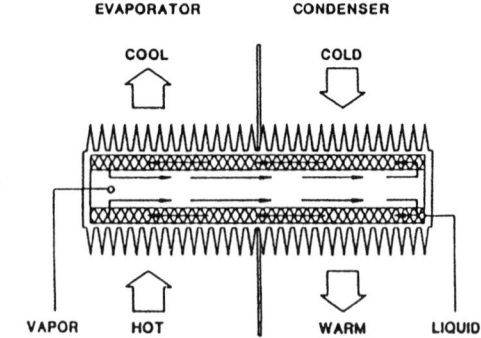
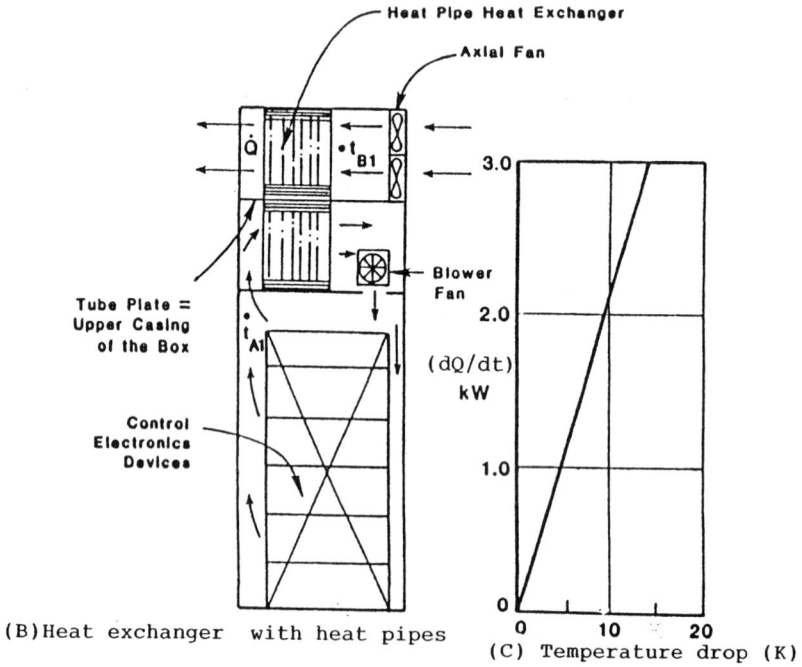

Fig. 8.3 Heat exchanger design using heat pipes showing (A) hot and cold streams, (B) cooling of "sealed" electronic package, and (C) heat flux rate as a function of temperature drop.

have extremely high thermal conductivities and possess uniform temperature distribution capability. Heat pipes are compact and lightweight and can eliminate localized hot spots at individual devices, printed circuit boards, or an entire system fabricated on a chip.

8.6 Impact of Material Properties on Cooler Performance

The effects of material properties and environments on the performance of cryogenically cooled microwave, MM-wave, infrared, and electro-optic components or sensors must be taken into account in the cooler design. The physical dimensions of these components or circuits can undergo radical changes that can adversely impact the performance levels of cryogenically cooled devices and circuits. The most serious and frequently observed operational problems under cryogenic operations include moisture condensation, freezing of waveguide or transmission lines leading to performance degradation in a microwave or MM-wave sensor, dimensional variations of critical components or devices with temperature, and variation in the thermal conductivity of various materials involved under cryogenic operations.

It is important to mention that both the linear thermal expansion and the thermal conductivity of various materials such as metals, alloys, and substrates vary under cryogenic operations and, thus, can adversely affect the overall performance of a device or sensor. Similarly, the effects of moisture in a cryogenically cooled device or system can severely degrade its performance and must be eliminated at all costs. Moisture can be eliminated in two ways: (1) by employing a "sealed" system where a moisture-free heat-exchanger gas or partial vacuum is present or (2) by allowing a moisture-free coolant to flow through the waveguide. According to the studies performed by the author, the latter method is inherently less reliable because of its susceptibility to moisture condensation if and when the coolant or refrigerant supply is exhausted. It is of paramount importance to mention that the heat-exchanger gas must remain in a gaseous state at the required cryogenic temperature. For example, dry helium gas could be used in a nitrogen-based cooling system; however, the nitrogen gas might liquefy in a helium-based refrigeration system [4].

8.6.1 PROPERTIES OF VARIOUS MATERIALS USED IN CRYOGENIC COOLERS

This section focuses on the thermal, electrical, and mechanical properties of various metals, alloys, and substrates that are frequently used in the design and development of cryogenic components and sensors. The adverse impact of these materials on the performance of various microwave, MM-wave, IR, and electrooptic (EO) devices or sensors under cryogenic environments is identified.

8.6.2 THERMAL PROPERTIES OF MATERIALS AT CRYOGENIC TEMPERATURES

The linear thermal coefficient of expansion, thermal conductivity, and the softening temperature of the materials used in the fabrication of a device or sensor at cryogenic operations are of critical importance, and the variations in these properties can affect the device performance. Prediction of a device's performance level at cryogenic temperature is strictly dependent on the type of metal or alloy and substrate used in its fabrication. Among the thermal properties, thermal conductivity is the most critical parameter of the material used in the fabrication of the cryogenically cooled device or component. However, among the dielectric materials, fused silica has the lowest value of the thermal expansion coefficient, while INVAR has the lowest among the alloys as illustrated in Figure 8.4 [4]. The widely used Teflon substrate in the microwave and MM-wave circuits and devices has the worst thermal expansion coefficient. The typical thermal expansion coefficient for the teflon is about 400×10^{-5} at 100 K and 280×10^{-5} at 77 K compared to 10×10^{-5} at 100 K and 4×10^{-5} at 77 K cryogenic temperature for the fused quartz hard substrate. Trade-off studies performed by the author reveal that fused quartz must be used for the substrate, INVAR for the Dewar housing, and copper for the conducting transmission lines or heat exchanger assembly to keep the dimensional variations to a minimum under cryogenic operations over the temperature range of 77 to 100 K.

Thermal conductivity of a material is another critical thermal parameter that indicates how fast it can transfer or remove the heat per unit thickness of the material at a given operating temperature. Thermal conductivity values of widely used materials in cryogenically cooled electronics, devices, or sensors are shown in Figure 8.5 [4]. Thermal studies

8. Requirements for Cryogenic Materials and Associated Accessories 239

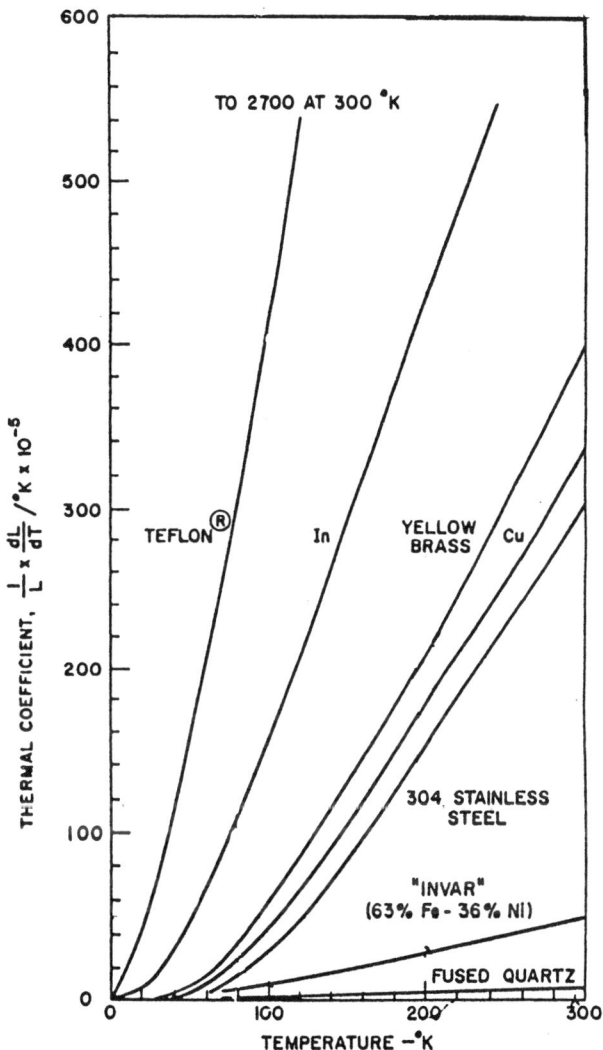

Fig. 8.4 Thermal coefficient as a function of temperature for various cooler materials.

performed by the author on various materials indicate that among the dielectric materials, teflon offers better thermal performance over fused quartz as illustrated in Figure 8.5. However, among the metals or alloys, pure copper offers the highest value of the thermal conductivity at cryogenic temperatures. It is evident from Figure 8.5 that the thermal

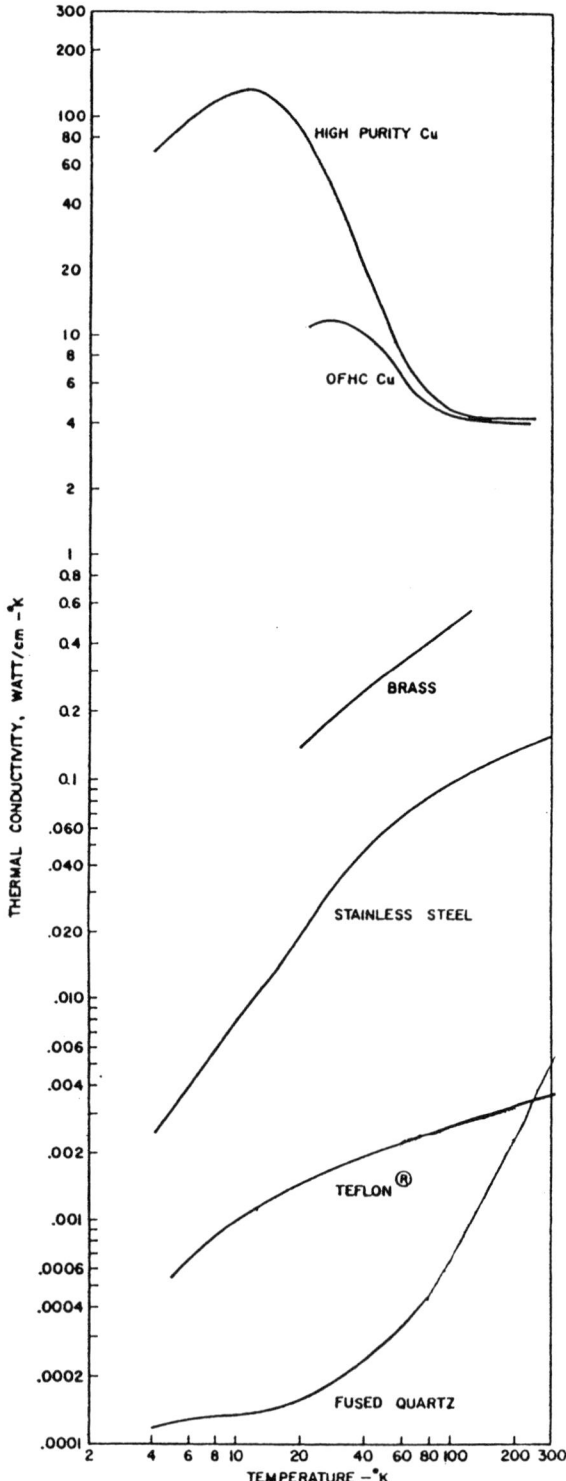

Fig. 8.5 Thermal conductivity as a function of temperature for various materials used in cryogenic coolers [1].

conductivity of teflon increases with the increasing cryogenic temperatures at a moderate temperature range of 4 to 20 K, but beyond this range it increases at a rapid rate over the 20 to 300 K temperature range.

8.6.3 ELECTRICAL PROPERTIES OF MATERIALS AT CRYOGENIC TEMPERATURES

Insertion loss, impedance mismatch, bandwidth limitation, and overall performance degradation in microwave and MM-wave devices or circuits depend on the electrical properties of the conducting and dielectric materials used. Dielectric losses in microwave and MM-wave devices depend on the permittivity, loss tangent, operating frequency, environmental temperature, thickness of the substrate, and the type of substrate (soft or hard or metallic) used. Most of the low-power microwave or MM-wave devices or components use soft or quasi-soft substrates such as Duroid 5880, 6002, or 6010 dielectric materials because of lower fabrication costs. However, cryogenic devices operating under severe mechanical and thermal environmental conditions may use hard substrates, such as sapphire, alumina, magnesium oxide, fused silica, and titanate ceramic compounds. In some applications, the anisotropic property and dimensional stability of the substrate materials are of paramount importance. Normalized permittivity, anisotropic values, and dimensional stability of Duroid 6010 substrate as a function of frequency and cryogenic temperature are shown in Figure 8.6. It is important to mention that the anisotropic parameter that is defined as the ratio of the dielectric constant or permittivity of the material along the xy-plane to that in z-axis has a serious impact on the device performance at cryogenic temperatures. Typical values of anisotropic parameters as a function of temperature are summarized in Table 8.1 as shown in Figure 8.6. Other important parameters, such as dielectric constant and loss tangent for soft substrates as a function of frequency and operating temperature, are shown in Figure 8.7.

As stated earlier, hard substrates such as strontium titanate, lanthanum aluminate, and magnesium oxide are best suited for cryogenic sensors operating under severe mechanical and thermal environments. Loss tangent and dielectric constants of some hard substrates as a function of cryogenic temperature at 10 GHz are depicted in Figure 8.8. It is evident from the curves shown in Figure 8.8(A) that the dielectric constant of these materials remains relatively constant over a wide temperature range, while the loss tangent values increase rapidly when the temperature falls

Typical Anisotropic Values for Various HTSC Soft Substrates

Substrate	e_{xy}/e_z	Dimensional Stability
Teflon (PTFE) ($e_r = 2.10$)	1.000	Excellent
Microfiber PTFE ($e_r = 2.20$)	1.025	Good
Microfiber PTFE ($e_r = 2.35$)	1.041	Fair
Woven PTFE ($e_r = 2.17$)	1.092	Poor
Woven PTFE ($e_r = 2.45$)	1.160	Worst
High dielectric Duroid ($e_r = 10.5$) with $t = 0.025$ in.	1.0201	Good
High dielectric Duroid ($e_r = 10.5$) with $t = 0.050$ in.	1.060	Fair
High dielectric Duroid ($e_r = 10.5$) with $t = 0.075$ in.	1.180	Worst

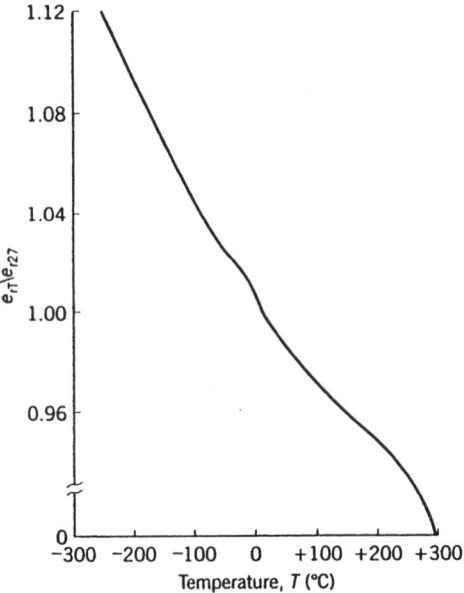

Fig. 8.6 Anisotropic property, dimensional stability, and normalized anisotropic value for soft substrates as a function of cryogenic temperature.

8. Requirements for Cryogenic Materials and Associated Accessories

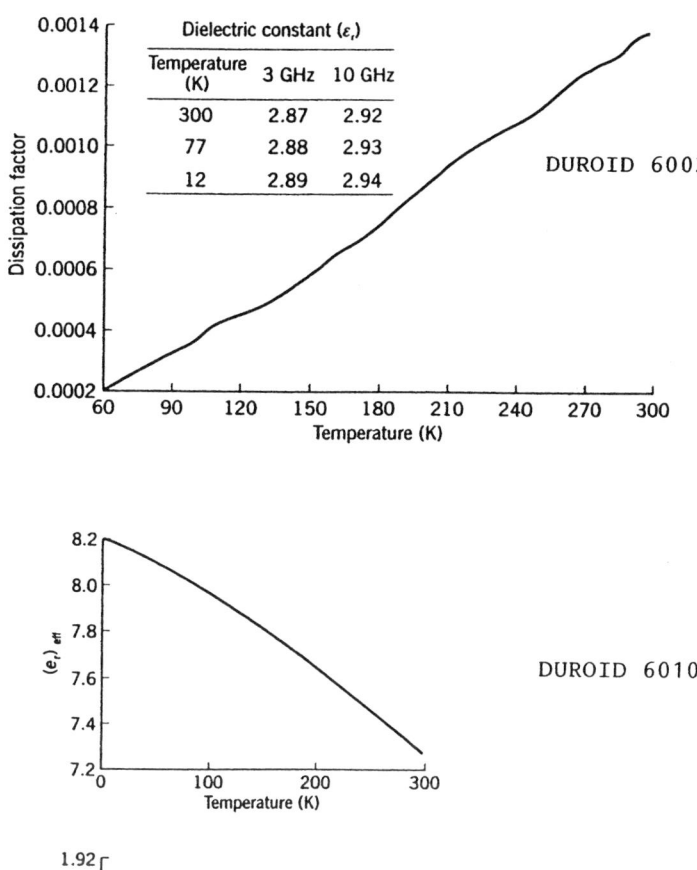

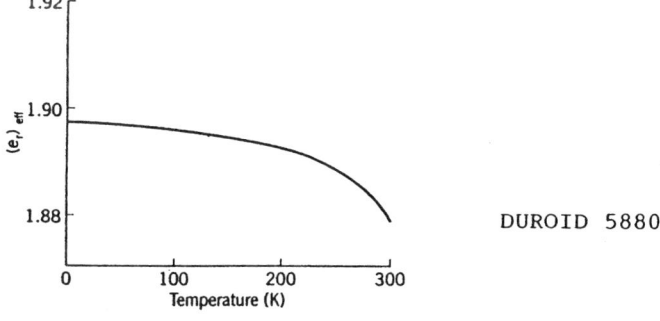

Fig. 8.7 Dissipation factor for Duroid 6002 and effective dielectric constant for Duroid 6010 and Duroid 5880 as a function of cryogenic temperature.

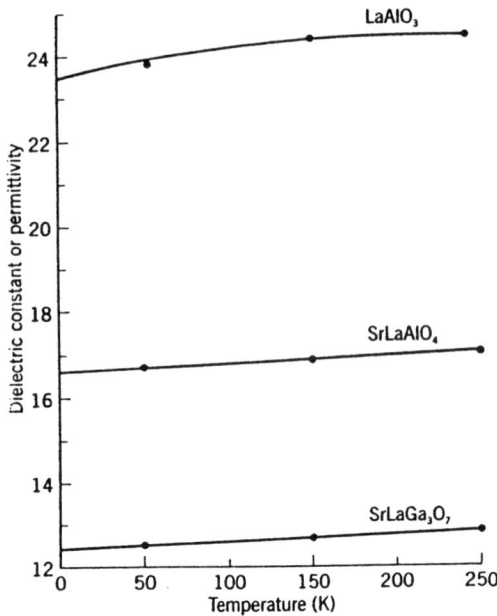

(A) Dielectric constant for superconducting substrates

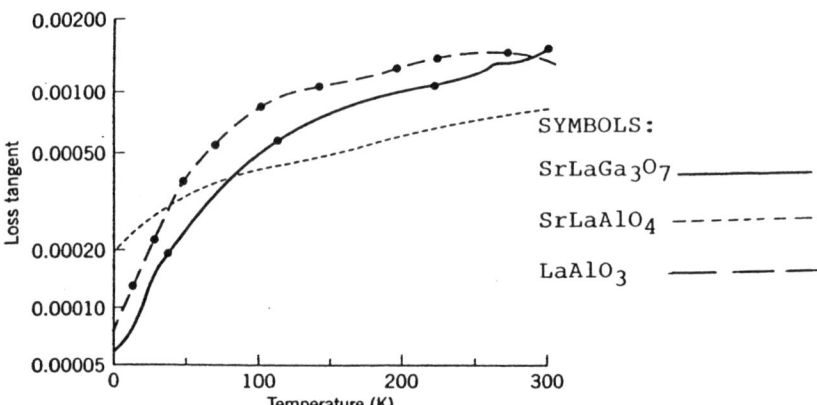

(B) Loss tangent values for superconducting substrates

Fig. 8.8 Dielectric constant and loss tangent for HTSC substrates as a function of cryogenic temperature.

8. Requirements for Cryogenic Materials and Associated Accessories

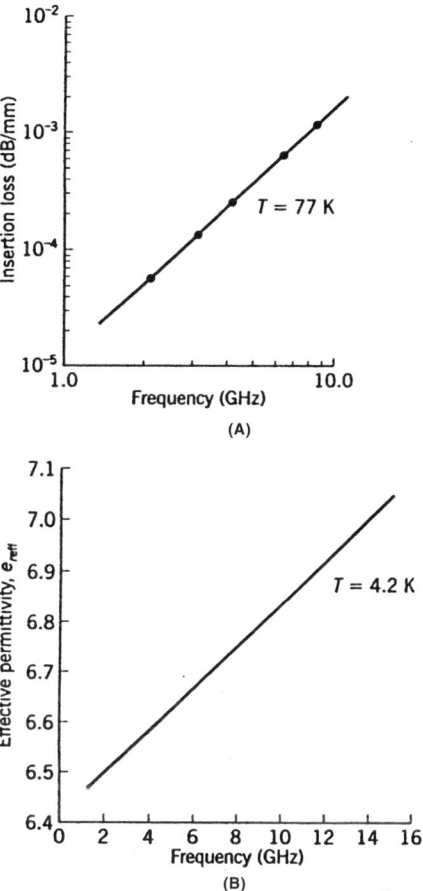

Fig. 8.9 Insertion loss (A) and effective permittivity (B) of the superconducting material MgO as a function of frequency and cryogenic temperature [5].

within the 4.2 to 100 K range. Because of the minimal cost for both the material and fabrication, magnesium oxide (MgO) substrate is widely used in the design and development of superconducting microwave and MM-wave components such as filters and low-noise RF amplifiers. Insertion loss and the effective dielectric constant of MgO substrate as a function of frequency and operating temperature [5] are shown in Figure 8.9. Note that both these parameters increase with the frequency and cryogenic temperature. However, insertion loss will be significantly lower at an operating temperature of 4.2 K, regardless of frequency. Fused silica substrate offers the lowest coefficient of thermal expansion and contraction, a minimum anisotropy parameter, and excellent dimensional stability over wide frequency and cryogenic temperature ranges.

8.6.4 MECHANICAL PROPERTIES OF MATERIALS AT CRYOGENIC TEMPERATURES

Mechanical properties of materials used in the fabrication of superconducting devices must be given serious consideration if the devices are required to operate under severe airborne and space environments. Important mechanical properties include tensile strength, compressive strength, and tensile modulus as a function of operating temperature. Their values as a function of cryogenic temperature are of critical importance while operating under severe mechanical environments. It is important to point out that in most alloys and metals, both the tensile strength and tensile modulus decrease as the operating temperature increases. However, in the case of fused silica substrate, their magnitudes increase with the increase in temperature as shown in Figure 8.10. This particular substrate [6] is best suited for microwave and MM-wave active and passive devices, where low insertion loss, excellent dimensional stability, and high mechanical strength under cryogenic operations are the principal requirements. Fused silica has demonstrated a tensile strength greater than 125,600 psi, a compressive strength better than 170,000 psi, tensile modulus of 10×10^6 psi, a shear modulus of 4.4×10^6 psi, and a modulus of rupture better than 5500 psi — all at an operating temperature of 77 K. This substrate offers the lowest values of coefficient of thermal expansion (CTE) and compression (CTC). It has a CTE of −0.6 micron per degree C at 77 K temperature.

The lowest values of CTE and CTC parameters are highly essential to prevent the solder and epoxy joints in the microwave devices or electronic circuits from cracking or becoming deformed under cryogenic operations. Furthermore, the magnitude of the CTC parameter for both a superconducting substrate and enclosure material must be not only low, but also must be as close as possible to each other to maintain a high RF performance level, improved reliability, and high mechanical integrity under cryogenic operations.

Metallic substrates are most ideal for HTSC high-power microwave devices and components. Thallium-based superconducting thick films are made from a TlBaCaCuO compound and when deposited on a silver alloy

8. Requirements for Cryogenic Materials and Associated Accessories

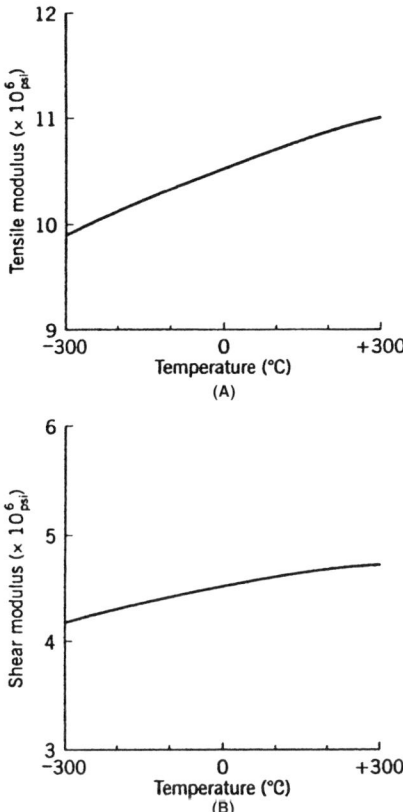

Fig. 8.10 Tensile modulus (A) and shear modulus (B) of hard substrate fused silica as a function of cryogenic temperature [2].

substrate using magnetron-sputtering process have demonstrated remarkable performance in superconducting high-power microwave cavities. These films exhibit the sharpest transition to the superconducting state at high frequency and demonstrate the weakest dependence of surface resistance on the magnetic field produced by high RF power levels. In summary, substrates for HTSC microwave or MM-wave devices and components must have low permittivity, a low loss tangent, unit anisotropy, high dimensional stability, fine grain structure, and matching CTE and CTC to provide an acceptable performance under severe mechanical and cryogenic environments.

8.7 Characteristics of Potential Refrigerants

Cooldown times depend on the type and amount of cryogen or refrigerant used. The cooling capacity, refrigeration efficiency, cooldown time, and amount of refrigerant used by the cryocooler determines the overall performance of a cooler. Frequent maintenance service is required irrespective of the cryogen used to keep the refrigeration system operating with high efficiency and stated reliability over extended durations. The cooling range of expendable cryogens that use the heat of vaporization vary from coolant to coolant. The limits of temperature ranges for an individual refrigerant are based on a minimum value as defined by the solid phase at 0.1 mm of mercury pressure and on a maximum value as defined by the critical point. Typical temperature ranges for selected expendable coolant are shown in Figure 8.11 [6]. Cryogenic cooling can be achieved at any operating temperature within the range of 2 to 300 K by selecting an appropriate refrigerant for specific environmental conditions. Critical parameters of potential cryogens are summarized in Table 8.3. Thermodynamic properties of potential liquid and gaseous cryogens are shown in Table 8.4.

Table 8.3 Important Characteristics of Potential Cryogens.

Cryogen	Boiling temperature (K)	Rel. vaporization	Specific power (W/W)
Helium	4.2	1*	1000
Nitrogen	77.0	64	30
Neon	27.2	41	140

*Indicates that 28 mw boils 1 liter of liquid helium per day.

Table 8.4 Thermodynamic Properties of Potential Liquid and Gaseous Cryogens.

Refrigerant	Boiling point (K)	Liquid den. (G/mL)	Gas den. (g/L)	Fusion heat (joules/g)	Critical temperature (K)
Helium	4.2	0.125	0.173	4.183	5.2
Nitrogen	77.4	0.808	1.250	25.52	126.1
Neon	27.2	1.200	0.910	16.72	44.5
Argon	87.4	1.399	1.780	28.05	150.8

8. Requirements for Cryogenic Materials and Associated Accessories

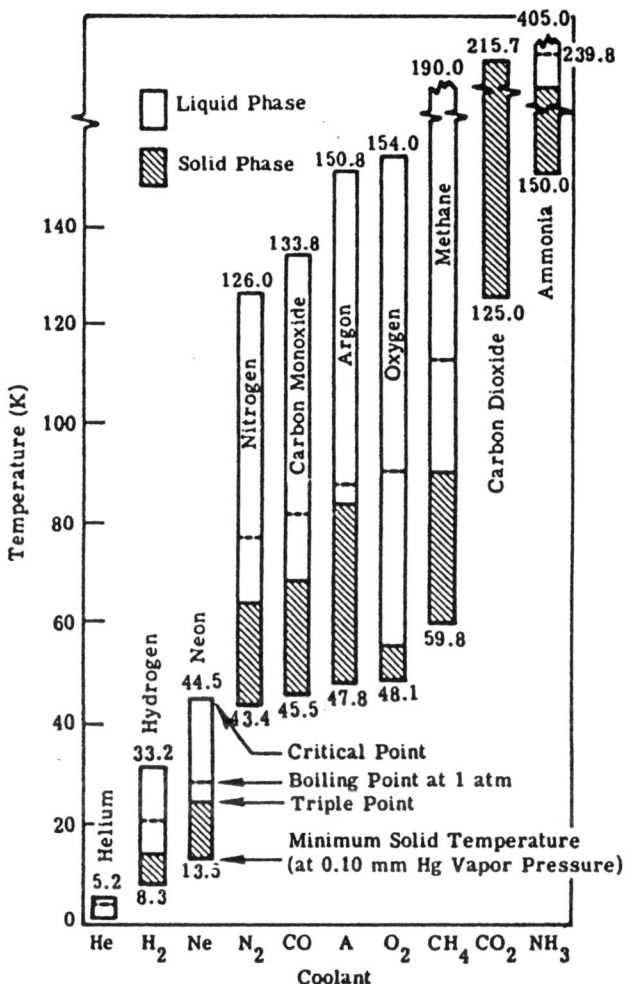

Fig. 8.11 Typical operating temperature ranges and boiling points for selected expendable coolants [3].

Studies performed by the author on various cryogens indicate that liquid cryogens are best suited for cooling complex airborne and space-based cryogenic systems. The logistic handling problems and high cost of liquid cryogen could be major disadvantages. To minimize these problems, current military cryogenic refrigerator systems deploy high-pressure gas, J-T-valve cryogenic coolers, and mechanical refrigeration systems.

8.7.1 COOLING CAPACITIES OF LIQUID CRYOGENS

Liquid refrigerants or cryogens are widely used in rocket propulsion systems for sophisticated long-range missiles, space tracking, and satellite communication applications. Cryogenic fluids may be stored as liquids in equilibrium with their vapors (subcritical) or at their higher pressures and temperatures as supercritical, homogeneous fluids. For scientific research and airborne applications, the cryogenic fluid is generally stored in the liquid state with a two-phase format because of the ease of operation and low storage weight. Note that in these applications, the fluid is constantly exposed to the supply port of the storage tank or Dewar assembly. For space storage of refrigerant or cryogen, the absence of gravity or acceleration from orientation forces prevents [6] the use of a standard two-phase system, because random orientation of the liquid phase during weightless environments prevents continuous communication between the liquid phase and the supply port. To ensure reliable communication, space-system storage of cryogenic liquids can be accomplished through pressurization of the cryogen to the supercritical pressure.

8.7.2 COOLING WITH SOLID CRYOGENS

A cooling system based on the sublimation of a solid cryogen into high vacuum of space avoids several operational problems associated with either subcritical or supercritical storage of liquid refrigerants. The operating temperature is dependent on the choice of solid cryogen and backpressure of the vent gas in the cooling system. The operating time or the cooldown time is strictly contingent on the amount of refrigerant deployed and the heat load. The heat of sublimation governs the cooling capacity of a system using solid cryogen, which is equal to the sum of the heat of vaporization and the heat of fusion. Solid cryogens offer several advantages, such as higher cooling capacity, independence of operating altitude, higher density storage, and lower operating temperatures. These unique qualities of a solid cryogen provide optimum cooling capability best suited for spaced-based IR sensors and long-range ballistic tracking systems requiring high rocket launching capability.

The operation of a refrigeration system using a solid cryogen is strictly based on the interaction of the pressure and temperature of a solid in equilibrium with its vapor. Operating pressure and temperature are maintained at constant levels through ventilation of the vapor to space at the designated pressure level [1]. Cryocoolers using solid refrigerants are normally designed without valves and the vent-gas ducting to maintain a specified operating pressure. Solid cryogens are best suited for rocket engines to be used in mobile missiles and second- and third-stage booster rockets to place the satellite payload in a geostationary orbit.

8.8 Maintenance Requirements for Various Cooler Accessories

Maintenance requirements for a cryocooler depend on the specific mission, cooling load, and cryogenic operating temperature. Accessories widely used in refrigeration systems — including G-M-closed coolers, Stirling-cycle coolers, immersion coolers, recirculating chillers, air-to-water heat exchanger based-coolers, and vapor-phase refrigeration (VPR) systems — are readily available. Furthermore, accessories with high mean time between failures (MTBF) or near-maintenance-free can be selected. Studies performed by the author on various cooling schemes indicate that the VPR system is still the most efficient, economical, commercially accepted refrigeration technology available to date, despite its major drawbacks including the wear and tear of moving parts and potential size limitations. It is the oldest means of mechanical refrigeration concept and is used extensively in domestic refrigerators, automobile air conditioning units, and central air conditioning systems in homes and commercial buildings. However, some cryogenic coolers used in airborne missiles and space-based sensors must have maintenance-free operation over 3 to 5 years.

8.8.1 *REQUIREMENTS FOR CRITICAL ACCESSORIES*

The most critical accessories in a refrigeration system are temperature controller, storage tank or Dewar assembly, and regulated power supply. Regulated power supplies are commercially available and can be procured to meet specific input and output voltage and current requirements. The power supply must have important features such as constant voltage amplitude regardless of load, overvoltage protection, automatic shutoff under overheating conditions, and the smallest current ripple amplitude.

Requirements for storage tanks or Dewar assemblies capable of storing gaseous, liquid, or sold cryogens were discussed in greater detail in Chapter 6. In the case of refrigerant storage tanks, special attention must be paid to heat loss, cryogen leakage, thermal shielding, and supply port location.

Temperature control is the most critical element of a refrigeration system. Precise temperature control and operational status of any refrigerator system are of paramount importance. Constant monitoring of refrigeration temperature requires three distinct controller functions to meet the specific cooler function: analog, digital, and microprocessor. The analog controller function offers full control over temperature parameters with high accuracy and reliability. The digital controller function provides remote sensing capabilities and direct instant communication with the computer. The microprocessor controller has the features of the digital controller and additional multistep programming capabilities such as adjustable high and low temperature safety settings and the ability to remove the controller for remote operation.

Note that a refrigerator bath/circulator offers a powerful refrigeration system incorporating a cycling stainless steel heater to provide precise temperature control and optimum performance stability. Oversized hermetic refrigeration compressor units must be permanently sealed for reliable, maintenance-free operation. An oversized compressor unit offers more cooling capacity, rapid temperature changes, and longer system life. Remote sensor performance monitoring capability requires both the microprocessor and the digital controller. The digital controller features a bright green light emitting diode (LED) display and allows the operator to select operating variables for temperature display. A refrigerator bath/circulator chiller can provide a temperature stability of $+/-0.01°C$ (digital) or $+/-0.1°C$ (analog), cooling capacity of 500 W at 20°C, and pumping capability of 15 liters/min. One can select the analog, digital, or microprocessor control function during the operation of the system. These bath refrigeration systems are compact and economical and are best suited for bench-top cooling applications or scientific research laboratories. Ultra-low-temperature bath/circulator systems offer an operating range of $-90°C$ to $-30°C$, temperature stability better than $+/-0.2°C$, cooling capacity of 340 W at $-80°C$, and pumping capacity of 16 liters/min at zero ft head.

8.8.2 CRYOGENIC INSULATION REQUIREMENTS

Cryogenic insulation is necessary for the chambers and pipes carrying the refrigerants. Multilayer insulation (MLI) composed of alternate layers of shield and spacer in high vacuum provides the most effective cryogenic insulation. MLI can be used for a cylindrical vessel, vacuum chamber, guard vessels, filling, and vent tubes. Accurate measurements of boiloff rate, temperature, interstitial pressure, and chamber pressure requires the MLI insulation. Thermocouples are located at appropriate places to measure the temperature distribution through the MLI thickness and to study the cooldown history. To reduce the moisture in the MLI medium, the MLI system is purged with nitrogen gas before and in between evacuations. MLI layers made from aluminized Mylar and glass fabric offer optimal insulation. MLI is dependent on the variations in effective thermal conductivity, which is a function on number of layers and layer density. Note that the heat flow rate decreases with an increase in the number of layers, whereas the heat flow rate increases with layer density.

8.8.3 IMPACT OF CRYOGENIC LEAK IN TUBES, FITTINGS, AND VALVES

Cryogenic leak in tubes, fittings, and valves must be avoided if reliable operation is the principal requirement. Cryogenic leak will decrease the cooling efficiency of the cooler and will require the cryogen to be filled often. Reliable leak detecting systems must be evaluated before deployment in a cryogenic system. Studies performed by the author indicate that a high-pressure and variable temperature cryogenic leak testing system will be most effective for testing leakage in pipe fittings and valves in liquid hydrogen supply lines. The fittings are tightened at room temperature (300 K). If the temperature of the fluid is high, the thermal expansion of the mating parts tends to tighten the seal and, hence, the leakage does not require any attention. If the fluid flows through a room temperature, tight-sealed fitting at cryogenic temperatures, such as liquid hydrogen or liquid helium, the mating parts of a conventional fitting would contract and a tight seal can no longer be assured. Preliminary calculations indicate that if the gap increases by 50% due to a temperature drop from 300 K to 77 K, a 337.5% increase in the leakage rate can be expected. The situation becomes more severe if the fluid is under high pressure.

8.8.4 IMPACT OF THERMOACOUSTIC OSCILLATIONS ON CRYOGENIC COOLERS

Thermoacoustic oscillations (TAOs) can develop in the connecting tubes between the cold and warm areas in the neck of tubes of Dewar vessels. TAOs in cryocoolers may cause high heat flux in the vessels, which can introduce errors in the parameter measurements. Furthermore, cryogenic piping should be dimensioned to satisfy the stability criteria. The tube or pipe has a constant temperature and radius outside the Dewar vessels. Stability limit can be determined for the system with different hot and cold pipe radii. The method described here allows approximate calculations for realistic temperature profiles as a function of pipe geometries. If the oscillations in the pipe geometry cannot be avoided, dampers have to be added in the system. Calculations can be made for the stability limit for various temperature profiles based on the principle that the gas in the tube acts as a free spring/mass oscillation. The motion of the cryogenic cooling system forces the gas through thermodynamic stability, which under certain conditions produces energy and drives the oscillations out. At the thermal stability limit, no power is produced. Proper design of cryogenic piping is necessary to avoid the TAOs. In brief, cryogenic piping should be dimensioned to satisfy the stability criteria. TAOs in the pipes can be calculated using monotonic temperature functions by integrating Rott's basic differential equations and using high-speed computers.

8.9 Summary

This chapter defined the requirements for various refrigeration systems, cryocooler materials, and critical accessories needed for system operation. Characteristics of important refrigerants or coolants were summarized with an emphasis on cooling efficiency and storage cost. Electrical, mechanical, and thermal properties of metallic, alloy, dielectric, and heat pipe materials were described. Potential problems frequently encountered in cryogenic cooling systems such as moisture condensation, heat loss, leakage of coolants, freezing of pipes carrying the refrigerants, and component dimensional changes with temperature variations were identified. Methods to improve cooler efficiency, reliability, and cooling capacity were discussed wherever possible. The impact of environmental effects

8. Requirements for Cryogenic Materials and Associated Accessories

on the cryocooler performance level was identified. Performance capabilities, limitations, cooling capacity, and the reliability of heat pipes were described in great detail. Performance capabilities and applications of TE coolers, immersion coolers, thermoacoustic coolers, magnetocaloric coolers, and vapor-phase refrigeration (VPR) systems were summarized, with a particular emphasis on heat-load capacity and cooling efficiency. Requirements of temperature controllers for various cooling applications were briefly discussed. Solid-state chillers used in the laser material process, therapeutic medicine, biomedical instruments, telecommunication products, microprocessor design, semiconductor processing, and aerospace and military device applications were briefly described. Reliability data on cryocoolers based on field operations sometimes are not readily available. However, published cryogenic literature indicates that commercial cryocoolers operating at a temperature of 5 K have demonstrated operating lifetimes of 15 years with standard 2-year maintenance.

References

1. W. L. Wolfe and G. J. Zissis. *The infrared handbook.* Environmental Research Institute of Michigan, Ann Arbor, MI: 1978, pp. 15–19.
2. W. L. Wolfe and G. J. Zissis. *The infrared handbook.* Environmental Research Institute of Michigan, Ann Arbor, MI: 1978, pp. 34–44.
3. G. P. Peterson. *An introduction to heat pipes.* New York: John Wiley & Sons, 1994, pp. 210–215.
4. T. P. Duprex. "Designing cryogenic microwave systems." *Microwaves*, October 1967, pp. 38–39.
5. A. R. Jha. *Superconductor technology: Applications to microwaves, electro-optics, electrical machines, and propulsion systems.* New York: John Wiley & Sons, 1998, p. 56.
6. W. L. Wolfe and G. J. Zissis. *The infrared handbook.* Environmental Research Institute of Michigan, Ann Arbor, MI: 1978, pp. 16–19.

Index

A

Active buffer stages, cooling efficiency with, 210–213, 213f
Adiabatic expansion, 2
Adiabatic expansion efficiency, 206, 208–210, 209t, 211f, 212f
Aerospace. *See* Space applications
Affordability, cryogenic technology, 6
Ammonia, 158f, 159t, 249f
Argon, 157t, 158f, 159t, 177t, 234, 248t, 249f
Availability, cryogenic technology, 7

B

Boltzmann, L., 60
Boreas cryocoolers, 104–106, 104f, 105f
  counterflow heat exchanger in design of, 180–182, 181f, 182f
Boreas-cycle cryocooler, 114
Boyle, Robert, 26
Brayton-cycle refrigeration systems, 146–147
  coefficient of performance for, 149–150, 150t
  performance characteristics of coolers with, 161, 162t
  reversed, 143–144
Buffer stages. *See* Active buffer stages

C

Carbon dioxide, 158f, 249f
Carbon monoxide, 158f

Carre, Ferdinand P. E., 26
CCC. *See* Closed-cycle cryogenic refrigerator
CCR. *See* Closed-cycle refrigeration
CCST. *See* Closed-cycle split-type Stirling cryocoolers
Chemical oxygen iodine laser (COIL), 191, 221
Chilled water storage system (CWS), 19–21, 22f
CHL. *See* Collins helium liquifier cryocooler
Claude-cycle refrigeration systems, 142–143, 143f
  performance characteristics of, 162t
Closed-cycle cryocoolers, 134
Closed-cycle cryogenic refrigerator (CCC), 114, 115f
Closed-cycle refrigeration (CCR)
  chronology of systems using, 100t
  cryogenic requirements for, 14–16, 14t
  cryopumping with, 17
  performance characteristics of coolers with, 160, 162t
Closed-cycle split-type Stirling cryocoolers (CCST), 173–176, 174f–176f
Coefficient of performance (COP)
  Brayton cooling cycle, 149–150, 150t
  refrigeration system, 147–150, 150t
  TE coolers, 231–232, 233f
Coefficient of thermal compression (CTC), 246–247
Coefficient of thermal expansion (CTE), 246–247
COIL. *See* Chemical oxygen iodine laser

257

Index

Cold storage (CS), requirements for, 13–14, 15f
Collins-cycle refrigerator systems, 120, 121f
Collins helium liquifier cryocooler (CHL), 167f, 168–170, 169f, 170t, 177
Composite wall, heat transfer in, 74f, 75–76
Conduction, heat transfer through, 52–53, 59–60, 59t
Convection
  heat transfer through, 53–54, 61–69, 62t, 63f, 65t–69t
  magnitudes for flow modes of, 62t
Cooldown time, 205–206, 207f, 208f
Cooler performance, material properties impact on, 237–247, 239f, 240f, 242f–245f, 247f
  electrical, 227t, 241–246, 242f–245f
  mechanical, 246–247, 247f
  thermal, 238–241, 239f, 240f
Cooler technologies, latest, 191–223
  advanced materials with, 191–192
  design concepts with, 191–192
  introduction to, 191
  pulse tube refrigerator as, 191–215, 193f, 196f, 198f, 200f–202f, 206f–208f, 209t, 211f–214f
    block diagram of, 193f
    coolant's impact on, 202–203, 202f
    design concepts for, 192–194, 193f
    double-orifice design configuration for, 193f, 194, 196f
    erbium-doped nickel-cobalt's impact on, 204
    heat leakage's impact on, 205–206, 207f, 208f
    operating parameter derivation for, 195–198, 196f, 198f
    parametric analysis of, 203–204
    performance capabilities of, 194–195
    performance improvement for, 199–206, 200f–202f, 206f–208f
    phase shift's impact on, 205, 206f
    regenerative material's impact on, 202–206, 202f, 206f–208f
    thermodynamic aspects of, 195–198, 196f, 198f
Cooling capacity, 5
  cryocoolers, 106–109, 107f, 108f
Cooling efficiency, 2
Cooling load capacity, 2
Cooling power, 208–210, 209t
COP. *See* Coefficient of performance
Coriolis effect, 43–44
Countercurrent flow, 90–91, 91t
Counterflow heat exchanger, 180–182, 181f, 182f
Cryo-guard, 11
Cryo-quick, 11
Cryo-trim, 11
Cryocoolers, 26–28
  active buffer stages for, 210–213, 213f
  adiabatic expansion efficiency with, 206, 208–210, 209t, 211f, 212f
  advanced technologies for integration in, 110–112, 111f
  Boreas, 104–106, 104f, 105f
  classifications of, 112–123, 115f–117f, 119f, 121f–123f, 134, 135t
  Boreas-cycle, 114
  closed-cycle, 114, 115f, 134, 135t, 173–176, 174f–176f
  Collins-cycle, 120, 121f
  Collins helium liquifier, 167f, 168–170, 169f, 170t, 177
  dilution/magnetic, 167f, 168
  G-M-cycle, 119–120, 170–171, 170t, 171f
  G-M-cycle cryocoolers employing J-T valves, 118–119, 119f, 170t, 171–172
  high-temperature, 120–123, 122f, 123f
  magnetic refrigerator systems, 230
  mechanical refrigerator, 229–230
  open-cycle, 134, 135t
  self-regulated J-T, 113, 173
  Stirling-cycle, 113, 170t, 172–173
  Stirling using advanced technologies, 114–118, 116f, 117f
  TE coolers, 230–232
  cooling power with, 208–210, 209t
  cooling schemes used by, 177–178, 177t, 178t
  counterflow heat exchanger in design of, 180–182, 181f, 182f
  critical design aspects of, 97–136
    cooling capacity in, 106–109, 107f, 108f

cooling power requirements in, 102–103
cyrogen storage requirements in, 133–134
cyrogen weight requirements in, 132–133
efficiency in, 129–132, 131f
high-pressure expansion ratio with, 106
input power requirements in, 99f
introduction to, 97
maintenance aspects in, 101–102, 102f, 167–168
mass flow rate optimization in, 109–110
operational requirements in, 97–98
performance requirements in, 98–104, 99f, 100t, 102f, 168–176, 169f, 170t, 171f, 175f, 176f
power estimates in, 127–129, 127t, 128f, 129t
pulse tube refrigeration in, 110–112, 111f
reliability requirements in, 101–102, 102f
specific weight in, 127–129, 127t, 128f, 129t
summary for, 134–136
temperature stabilization in, 109–110
thermodynamic, 129–132, 131f
critical thermodynamic aspects of, 183
design concepts/materials for, 179–182, 181f, 182f
high-power laser systems' requirements for, 220–221
high-pressure ratio in design of, 180–182, 181f, 182f
high-pressure ratios used in, 104–106, 104f, 105f
higher cryogenic temperatures with, 229–232
industrial applications with, 206–210, 209t
IR missiles' requirements for, 228–229
IR sensors' requirements for, 216–220, 217f
low-temperature switch with, 206–207, 210
materials/accessories needed for, 225–255

cooler performance from, 237–247, 239f, 240f, 242f–245f, 247f
electrical properties of, 227t, 241–246, 242f–245f
heat pipes as, 232–237, 234f, 236f
insulation requirement with, 253
introduction to, 225–226
leaks impact on, 253
maintenance requirements of, 251–254
mean time between failures of, 251
mechanical properties of, 246–247, 247f
potential refrigerants as, 248–251, 248t, 249f
properties of, 238
summary of, 254–255
temperature control of, 252
thermal properties of, 238–241, 239f, 240f
thermoacoustic oscillations impact with, 254
medical applications' requirements for, 222–223
microwave's requirements for, 215–216
military applications' requirements for, 227–229, 228t
minimum refrigeration temperature of, 199, 200f
multibypass technique for, 210–214, 214f
optimizing cooling capacity in, 184–185, 184f, 186f
optimizing mass flow rate in, 185–187
optimizing temperature stability in, 185–187
rare-earth materials in design of, 179–180
requirements for, 165–190
  EO sensors, 165
  focal planar arrays, 165, 166
  introduction to, 165
  IR sensors, 165, 166, 170t
  MRI, 166, 170, 170t, 172
  summary of, 189–190
scientific research requirements for, 229
sonar applications' requirements for, 221–222

Cryocoolers (*continued*)
  space applications design requirements for, 187–189, 188f, 189t, 226–227, 227t
  Stirling-cycle, 102, 113
  surveillance imaging requirements for, 226–227, 227t
  technology developments of, 26–27
  thermal losses for two-stage, 178, 178t
  thermodynamic aspects of, 27–28, 28t
  thermodynamic efficiency limits with, 183
Cryoelectronics, 5
Cryogen, 2, 5
  characteristics of, 157–158, 157t, 158f, 248–251, 248t, 249f
  cooling systems configurations using, 156–160, 157t, 158f, 159f
  gases
    important properties of, 27–28, 34f, 248–249, 248t, 249f
    performance characteristics of coolers with, 162t
  liquid, 2, 250
  solid, 158–159, 159t, 160, 162t, 250–251
  storage requirements in cryocoolers for, 133–134
  techniques to reduce leak with, 160
  weight requirements in cryocoolers for, 132–133
Cryogenic Dewar, 151–154, 152t, 153f
Cryogenic technology
  aerospace application with, 8–9
  affordability of, 6
  availability of, 7
  benefits from, 6–7
  chilled water storage system with, 19–21, 22f
  cold storage with, 13–14, 15f
  component performance improvements with, 7
  critical aspects/issues in, 5–6
  development history of, 1–22
  early applications of, 8–9
  frozen food industry with, 13
  gas production with, 8
  gas separation process with, 10
  ice-making machines with, 18–19, 20f, 21f
  ice-storage systems with, 18–19, 18f, 19f
  industrial applications of, 10–12, 16–21, 18f–21f
  line regulators with, 9
  liquid level controller with, 9
  medical applications with, 14–16, 14t
  MRI with, 14, 16
  nuclear radiation testing with, 17–18
  prominent contributors to, 2–5, 3f
  summary of, 21–22
  terms associated with, 2
Cryogenic temperature, 2
Cryopumping, 17
CS. *See* Cold storage
CTC. *See* Coefficient of thermal compression
CTE. *See* Coefficient of thermal expansion
Cullen, Dr. William, 26
CWS. *See* Chilled water storage system

D

Dewas, 5
Dilution-magnetic cryocoolers, 167f, 168
Dilution refrigeration systems, 229
Double-orifice design, pulse tube refrigerator configuration with, 193f, 194, 196f

E

Electrical properties of cooler materials, 227t, 241–246, 242f–245f
  anisotropic parameters in, 241, 242f
Electrooptic sensors (EO sensors), cryocoolers requirements for, 165
Enthalpy, symbol for, 28t
Enthalpy flow, 198f
Entropy, symbol for, 28t
EO sensors. *See* Electrooptic sensors
Erbium-doped nickel-cobalt (ErNi), 204
ErNi. *See* Erbium-doped nickel-cobalt
Ethane, 234

F

Faraday, Michael, 1, 2
FCR. *See* Frictional coefficient ratio
FDA. *See* Federal Drug Administration
Federal Drug Administration (FDA), frozen food industry requirements from, 13
Figure of merit (FOM)
 heat pipes with, 234, 234f
 TE coolers, 231
FLIR. *See* Forward-looking IR sensors
Fluid flows, convection with, 61, 63–64
Fluids, heat transfer coefficients for, 82t
Focal planar array (FPA), 123
 cryocooler requirements for, 216
 cryocoolers requirements for, 165, 166
FOM. *See* Figure of merit
Formulas
 thermodynamic, 27, 29t–31t
 viscosity conversion in, 27, 32t–33t
Forward-looking IR sensors (FLIR), 126
FPA. *See* Focal planar array
Freon, 234
Frictional coefficient ratio (FCR), 34–41, 34f–39f
 rotational Reynolds number with, 34, 40
 rotation's impact on, 39, 39f
 simple calculation showing, 35
Frozen food industry, 13

G

Gas liquefaction, 1, 2, 8
Gas separation process, 10
Gases. *See also* Cryogen; Liquefied gas
 ambient specific heat for, 87t
 production of, 8
Gibbs' free-energy, 195
 symbol for, 28t
Gifford-McMahan cycle cryocoolers (GM cycle cryocoolers), 3–4, 3f, 14, 98, 119–120, 144–146, 145t
 employing J-T valves, 118–119, 119f, 170t, 171–172
 history of, 3–4, 3f
 performance characteristics of, 160–161, 162t, 170–171, 170t, 171f
Gorrie, John, 4, 26
Gorter-Mellink effects, 48–50
GT-cycle, 2

H

HCRS. *See* High-capacity refrigeration systems
Heat capacity, commercial refrigerants, 12, 12t
Heat energy flow, 198f
Heat exchanger pipes
 cylindrical, 76–78, 77f
 heat transfer through, 76–79, 77f
 insulated, 77f, 78–79
Heat exchangers
 composite wall with, 74f, 75–76
 conduction mode of, 52–53
 convection mode of, 53–54
 counterflow, 180–182, 181f, 182f
 critical parameters for, 89–93, 90t, 91t
  clean conditions as, 92–93
  countercurrent flow as, 90–91, 91t
  fouling conditions as, 92–93
  number of exchanger tubes as, 93–94
  outside exchanger surface area as, 91–92
  shell diameter as, 93
 design aspects for, 79–85, 80f, 81t, 82t, 84f, 85f, 86t
 efficient tube/rod layout for, 80, 80f
 fluid temperature distributions in, 85f
 heat load calculations for, 79–85, 80f, 81t, 82t, 84f, 85f, 86t
 heat pipes with, 236f
 heat transfer coefficients for, 81t, 82t
 heat transfer modes with, 71–76, 74f
 linear heat flow's impact on, 40–41, 42f
 planar wall with, 71–75, 74f
 preliminary rating of, 94
 radiation mode of, 54–55
 shell-and-tube, 81t
 transfer rates for, 52–55
 turbulent flow's impact on, 41–47, 44f–47f

Index

Heat flows, 25–55
  convection with, 61
  Coriolis effect with, 43–44
  cryogenic coolers with, 26–28
    technology developments of, 26–27
    thermodynamic aspects of, 27–28, 28t
  Gorter-Mellink effects with, 48–50
  heat exchangers with, 52–55
  introduction to, 25
  linear, 33–41, 34f–39f
  Nusselt number with, 43, 45f–47f, 46
  rotation with, 33–34, 34f, 39f, 40–47, 41f, 44f–47f
  summary of, 55
  transfer rates with, 52–55
  turbulent flow in, 41–47, 44f–47f
  two-dimensional models of, 47–52, 51f
  types of, 28, 33–47, 34f–39f, 41f, 42f, 44f–47f
Heat leakage, 205–206, 207f, 208f
Heat load, calculations of, 79–85, 80f, 81t, 82t, 84f, 85f, 86t
Heat load capacity, 2
Heat pipes. *See also* Heat exchanger pipes
  concept, 232–237, 234f, 236f
  figure of merit with, 234, 234f
  heat exchanger design using, 236f
  performance limitations with, 234–237, 234f, 236f
Heat-powered refrigerator cooling system, 130–132, 131f
Heat transfer, 52–55
  composite wall with, 74f, 75–76
  computation of overall coefficient of, 86–89, 87t
    clean environment with, 89, 90t
    foul conditions with, 88–89
  conduction mode of, 52–53, 59–60, 59t
  convection mode of, 53–54, 61–69, 62t, 63f, 65t–69t
  cylindrical pipes with, 76–78, 77f
  heat exchanger performance with, 71–76, 74f
  heat exchanger pipes with, 76–79, 77f
  insulated pipes with, 77f, 78–79
  modes of, 57–58, 59–71, 59t, 61t, 62t, 63f, 65t–69t, 71t
  numerical example of, 69–71, 71t, 72f, 73f
  planar wall with, 71–75, 74f
  radiation mode of, 54–55, 60–61, 61t
  thermodynamic aspects of, 57–95, 59t, 61t, 62t, 63f, 65t–69t, 71t, 72f–74f, 77f, 80f, 81t–82t, 84f, 85f, 86t, 90t, 91t
  three laws of, 58–59
  wall thickness with, 59t
Helium, 157t, 158f, 177, 177t, 234, 248t, 249f. *See also* Liquid helium
Helmholtz free-energy, 28t
High-capacity coolers
  heat exchanger design aspects for, 79–85, 80f, 81t, 82t, 84f, 85f, 86t
  heat load calculations for, 79–85, 80f, 81t, 82t, 84f, 85f, 86t
  thermodynamic aspects of, 57–95, 59t, 61t, 62t, 63f, 65t–69t, 71t, 72f–74f, 77f, 80f, 81t–82t, 84f, 85f, 86t, 90t, 91t
  cylindrical pipes with, 76–78, 77f
  heat exchanger performance with, 71–76, 74f
  heat exchanger pipes with, 76–79, 77f
  heat transfer conduction with, 52–53, 59–60, 59t
  heat transfer convection with, 53–54, 61–69, 62t, 63f, 65t–69t
  heat transfer critical parameters with, 89–93, 90t, 91t
  heat transfer modes with, 57–58, 59–71, 59t, 61t, 62t, 63f, 65t–69t, 71t
  heat transfer numerical example with, 69–71, 71t, 72f, 73f
  heat transfer phenomenon with, 57–58
  heat transfer preliminary rating with, 94
  heat transfer radiation with, 54–55, 60–61, 61t
  heat transfer three laws with, 58–59
  insulated pipes with, 77f, 78–79
  introduction to, 57
  summary of, 94–95
High-capacity refrigeration systems (HCRS), 139–142, 140f–142f
High-power laser systems, 220–221

High-pressure ratio, cryocoolers design
    with, 180–182, 181f, 182f
High-temperature refrigerator systems,
    120–123, 122f, 123f
  cooling power requirements with
    superconducting, 122–123, 123f
High-temperature superconductor
    technology (HTSC), 102, 120,
    122, 221
Hydrogen, 157t, 158f, 159t

I

Ice-making machines, 18–19, 20f, 21f
Ice-storage systems, 18–19, 18f, 19f
Industrial applications, of cryogenic
    technology, 16–21, 18f–21f
  chilled water storage system as, 19–21,
    22f
  cryocooler designs for, 206–210, 209t
  cryopumping as, 17
  ice-making machines as, 18–19, 20f,
    21f
  ice-storage systems as, 18–19, 18f, 19f
  nuclear radiation testing as, 17–18
Infrared sensors (IR sensors), 126
  cryocooler requirements for, 165, 166,
    170t, 216–220, 217f
  PTR with, 193
Insulation requirements, 253
Internal energy, symbol for, 28t
IR countermeasures (IRCM), 219
IR missile warning receiver (IRMW), 219
IR missiles, tactical coolers for, 228–229
IR search and tracking (IRST), 219
IR sensors. *See* Infrared sensors
IRCM. *See* IR countermeasures
IRMW. *See* IR missile warning receiver
Irreversible adiabatic expansion, 2
IRST. *See* IR search and tracking
Isenthalpic expansion, 2
Isentropic expansion, 2

J

J-T closed cycle, 5
J-T cryocooler. *See* Joule-Thomson
    cryocooler

J-T effect. *See* Joule-Thomson effect
J-T expansion valve, 2, 14
J-T heat exchanger, 5
Joule, 2
Joule-Thomson cryocooler (J-T
    cryocooler), 146, 146t
  performance characteristics of, 127t,
    160, 162t
  self-regulated, 113
Joule-Thomson effect (J-T effect), 4, 98

K

Kirk cycle, 3–4, 3f
Krypton, 234

L

Laminar flow
  convection with, 61, 64
  rotation with, 40, 41f
Leaks in tubes/fittings/valves, impact of,
    253
LEO. *See* Low-earth orbiting satellite
Leslie, Sir John, 26
LIN. *See* Liquid nitrogen
Linde, Carl P. G., 26
Linde cycle, 4
Line regulators, 9
Linear heat flow, 33–41, 34f–39f
  heat exchanger performance's impact
    on, 40–41, 42f
Liquefied gas
  performance characteristics of coolers
    with, 162t
  temperature/pressure requirements for
    storing, 142f, 156
Liquid-feed cooling system, 155, 155f
Liquid helium, 1, 4–5, 5
  properties of, 12t
  superfluid, 47, 49, 51f, 52
Liquid hydrogen, 11, 12t
Liquid level controller (LLC), 9
Liquid neon, 11, 12t
Liquid nitrogen (LIN), 11–12, 12t
Low-earth orbiting satellite (LEO),
    187
Low-temperature physic theory, 5

Low-temperature switch (LTS), cryocooler design with, 206–207, 210

M

MAC. See Maximum cooling capacity
MACC. See Maximum available cooling capacity
Magnetic refrigerator systems (MRS), 230
Magnetic resonance imaging (MRI), 1, 5, 98, 120
  cryogenic requirements for, 14, 16, 166, 170, 170t, 172, 222
  PTR with, 193
Magnetohydrodynamic systems (MHD), 169, 170t
Maintenance Breelife, 5
Mass flow rate, 185–187
  optimization, 109–110
Maximum available cooling capacity (MACC), 106
Maximum cooling capacity (MAC), 184–185, 184f
Mean time between failures (MTBF), 251
Mechanical-powered refrigerator cooling system, 130–132, 131f
Mechanical properties of cooler materials, 246–247, 247f
Mechanical refrigerator (MR), 229–230
Medical applications
  cryocooler requirements for, 222–223
  cryogenic requirements for, 14–16, 14t
MET. See Minimum refrigeration temperature
Metals
  ambient specific heat for, 87t
  thermal conductivity of, 87t
Methane, 158f, 234, 249f
MHD. See Magnetohydrodynamic systems
Microcoolers. See also Cryocoolers
  cooling schemes used by, 177–178, 177t, 178t
  critical design aspects of, 97–136
    cooling power requirements in, 103–104
    introduction to, 97
    mass flow rate optimization in, 109–110

operational requirements in, 97–98
performance capabilities with, 123–126, 124t, 125t
summary for, 134–136
temperature stabilization in, 109–110
cryogens for, 124, 124t
focal planar array with, 123
limitations of, 126
military applications requirements for, 178–179, 180f
optimizing cooling capacity in, 184–185, 184f, 186f
optimizing mass flow rate in, 185–187
optimizing temperature stability in, 185–187
potential cooling schemes for, 124–125, 124t, 125t
rare-earth materials in design of, 179–180
requirements for, 165–190
  introduction to, 165
  summary of, 189–190
space applications requirements for, 178–179, 180f
typical thermal losses for, 125, 125t
Microwave, 215–216
Military
  cryocooler requirements for, 227–229, 228t
  IR missiles for, 228–229
  microcoolers requirements for, 178–179, 180f
  operational requirements in, 97–98
Minimum refrigeration temperature (MET), 199, 200f
Missile applications
  liquid-feed requirements for, 155, 155f
  storage tank requirements for, 154–156, 155f, 156t
  transfer line requirements for, 155–156, 156t
MLI. See Multilayer insulation
MM-wave systems, 215–216
Molar enthalpy, 195
Monroe, Charles E., 26
MR. See Mechanical refrigerator
MRI. See Magnetic resonance imaging
MRS. See Magnetic refrigerator systems
MTBF. See Mean time between failures

Multibypass technique, 210–214, 214f
Multilayer insulation (MLI), 253

N

NAR. *See* Nose area ratio
Neon, 158f, 177t, 248t, 249f
Nitrogen, 157t, 158f, 159t, 177t, 248t, 249f
Nose area ratio (NAR), 187
Nuclear radiation testing, 17–18
Nusselt number, 43, 45f–47f, 46
  convection with, 63–64
  heat transfer numerical example with, 69–71, 71t, 73f

O

Open-cycle cryocoolers, 134
Oxygen, 158f, 234, 249f

P

Perkins, Jacob, 26
Phase shift, cooling performance base on, 205, 206f
Planar wall, heat transfer in, 71–75, 74f
Prandtl number, 33, 43
  convection with, 63–64
  heat transfer numerical example with, 69–71, 71t, 73f
Pressure, symbol for, 28t
Pseudo-Reynolds number, 33, 34, 41
Pulse tube refrigerator (PTR), 191–215, 193f, 196f, 198f, 200f–202f, 206f–208f, 209t, 211f–214f
  active buffer stages for, 210–213, 213f
  adiabatic expansion efficiency with, 206, 208–210, 209t, 211f, 212f
  block diagram of, 193f
  coolant's impact on, 202–203, 202f
  cooling power with, 208–210, 209t
  design concepts for, 192–194, 193f
  double-orifice design configuration for, 193f, 194, 196f
  erbium-doped nickel-cobalt's impact on, 204
  heat leakage's impact on, 205–206, 207f, 208f
  low-temperature switch with, 206–207, 210
  multibypass technique for, 210–214, 214f
  operating parameter derivation for, 195–198, 196f, 198f
  parametric analysis of, 203–204
  performance capabilities of, 194–195
  performance improvement for, 199–206, 200f–202f, 206f–208f
  phase shift's impact on, 205, 206f
  regenerative material's impact on, 202–206, 202f, 206f–208f
  system design aspects with, 110–112, 111f
  thermodynamic aspects of, 195–198, 196f, 198f

R

Radiation, heat transfer through, 2, 54–55, 60–61, 61t
Radiation testing, 17–18
Rare-earth materials (RE), cryocoolers design with, 179–180
Refrigerants, 248–251, 248t, 249f. *See also* Cryogen
Refrigerants, commercial
  heat capacity of, 12, 12t
  temperature ranges for, 167
Refrigeration systems
  Brayton-cycle, 146–147, 149–150, 150t, 161, 162t
  Claude-cycle, 142–143, 143f, 162t
  coefficient of performance for, 147–150, 150t
  Collins helium liquifier, 167f, 168–170, 169f, 170t, 177
  cryogenic Dewar requirements for, 151–154, 152t, 153f
  cryogens/cooling agents for, 156–160, 157t, 158f, 159f
  description of, 139–142, 140f–142f
  dilution/magnetic, 167f
  G-M-cycle, 119–120, 144–146, 145t, 167f, 170–171, 170t, 171f

Refrigeration systems (*continued*)
  G-M-cycle employing J-T valves, 118–119, 119f, 170t, 171–172
  high-capacity, 139–142, 140f–142f
  J-T-cycle, 146, 146t
  liquefied gas storage requirements with, 142f, 156
  moderate cooling capacity, 144–147, 145t, 146t
  performance comparison of, 160–161, 162t
  performance requirements for, 139–162
  reversed-Brayton-cycle, 143–144
  space/missile storage tank requirements for, 154–156, 155f, 156t
  Stirling-cycle, 114–118, 116f, 117f, 127, 128f, 129f, 160–161, 162t, 167f, 170t, 172–173
  storage tank requirements for, 151–156, 152t, 153f, 155f, 156t
  thermo-electric, 167f
  turbo-machinery, 147, 148f, 149f
  vapor compression, 167f
Reversed-Brayton-cycle refrigeration systems, 143–144
Reynolds number, 33, 34, 40, 43
  convection with, 63
  heat transfer numerical example with, 69–71, 71t, 72f, 73fr
Rotating system geometry, 33, 34f
Rotation
  Coriolis effect with, 43–44
  flow resistance ratio's impact on, 39, 39f
  geometry of, 33–34, 34f
  heat flows with, 33–34, 34f, 39f, 40–47, 41f, 44f–47f
  laminar flow resistance's impact on, 40, 41f
  orthogonal-mode, 41–43
  parallel-mode, 41
  turbulent flows with, 41–47, 41f, 44f–47f

S

Scientific research, dilution refrigeration systems for, 229
Shell-and-tube heat exchanger, 82t
Shell diameter, 93
Solid cryogens, 158–159, 159t, 250–251
  performance characteristics of coolers with, 160, 162t
Sonar applications, 221–222
Space applications
  cryocoolers design requirements for, 187–189, 188f, 189t, 226–227, 227t
  cryogenic technology for, 5, 8–9
  front-end component requirements for, 226–227, 227t
  liquid-feed requirements for, 155, 155f
  microcoolers requirements for, 178–179, 180f
  operational requirements in, 97
  storage tank requirements for, 154–156, 155f, 156t
  transfer line requirements for, 155–156, 156t
Specific weight, 127–129, 127t, 128f, 129t
SQUID. *See* Superconducting quantum interference device
Stefan, J., 60
Stirling-cycle cryocoolers, 113
  advanced technologies used by, 114–118, 116f, 117f
  IR sensors' requirements for, 216–220, 217f
  performance characteristics of, 127, 129f, 160–161, 162t, 170t, 172–173
  weight/power characteristics of, 127, 128f
Stirling cycles, 3, 3f
Storage tank
  cryogenic Dewar requirements for, 151–154, 152t, 153f
  refrigeration systems requirements for, 151–156, 152t, 153f, 155f, 156t
  space/missile requirements for, 154–156, 155f, 156t
Superconducting quantum interference device (SQUID), 110, 114, 118, 170t, 172, 222, 226
  PTR with, 193
Superfluid helium, two-dimensional heat flow model with, 47, 49, 51f, 52
Surveillance imaging, 226–227, 227t
Symbols, thermodynamic, 27, 28t

T

TAO. *See* Thermoacoustic oscillations
TBE. *See* Thermal-balance equations
TE coolers. *See* Thermoelectric coolers
Temperature, symbol for, 28t
Temperature stability, 185–187
Temperature stabilization, 109–110, 252
Thermal-balance equations (TBE), 188–189, 189t
Thermal coefficient, 238, 239f
Thermal conductivity, 238–241, 240f
Thermal-electric coolers (TE coolers), 110
Thermal properties of cooler materials, 238–241, 239f, 240f
Thermo-electric refrigeration systems, 167f
Thermoacoustic oscillations (TAO), 254
Thermodynamics
  analysis of refrigeration with, 130–132, 131f
  cooling cycles with, 183
  critical aspects of cryocoolers in, 183
  cryogenic coolers with, 27–28, 28t
  design of cryocoolers with, 129–132, 131f
  formulas used in, 27, 29t–31t
  heat exchanger design aspects with, 79–85, 80f, 81t, 82t, 84f, 85f, 86t
  heat load calculations with, 79–85, 80f, 81t, 82t, 84f, 85f, 86t
  heat transfer in, 57–95, 59t, 61t, 62t, 63f, 65t–69t, 71t, 72f–74f, 77f, 80f, 81t–82t, 84f, 85f, 86t, 90t, 91t
  composite wall with, 74f, 75–76
  computation of overall coefficient of, 86–89, 87f, 90t
  conduction mode of, 52–53, 59–60, 59t
  convection mode of, 53–54, 61–69, 62t, 63f, 65t–69t
  cylindrical pipes with, 76–78, 77f
  heat exchanger performance with, 71–76, 74f
  heat exchanger pipes with, 76–79, 77f
  insulated pipes with, 77f, 78–79
  modes of, 57–58, 59–71, 59t, 61t, 62t, 63f, 65t–69t, 71t
  numerical example of, 69–71, 71t, 72f, 73f
  planar wall with, 71–75, 74f
  radiation mode of, 54–55, 60–61, 61t
  three laws of, 58–59
  linear heat flow in, 33–41, 34f–39f
  pulse tube refrigerator, 195–198, 196f, 198f
  symbols used in, 27, 28t
  viscosity conversion in, 27, 32t–33t, 64–69, 65t–68t
Thermoelectric coolers (TE coolers), 230–232
  COP for, 231–232, 233f
  FOM for, 231, 234f
  junction temperature for, 233f
  load capacity for, 233f
  performance parameters of, 231–232
Total heat, symbol for, 28t
Trains, magnetically levitated, 169, 170t
Transfer lines, 155–156, 156t
Tudor, Frederic, 26
Turbo-machinery refrigeration systems, 147, 148f, 149f
Turbulent flow
  convection with, 61, 64
  heat exchanger performance's impact on, 41–47, 44f–47f

V

Vapor compression refrigeration systems, 167f
Vapor-phase refrigeration (VPR), 251
Viscosity, 64–69, 65t–68t
  conversion, 27, 32t–33t
  conversion factors, 65t–67t
  of gases, 68t
  of liquids, 68t
  temperature with, 68t
Volume, symbol for, 28t
VPR. *See* Vapor-phase refrigeration

W

Wind tunnel applications, 5, 8–9